LOGIC AND ARITHMETIC

Logic and Arithmetic

* Natural Numbers

DAVID BOSTOCK

OXFORD · AT THE CLARENDON PRESS
1974

Oxford University Press, Ely House, London W. 1

GLASGOW NEW YORK TORONTO MELBOURNE WELLINGTON
CAPE TOWN IBADAN NAIROBI DAR ES SALAAM LUSAKA ADDIS ABABA
DELHI BOMBAY CALCUTTA MADRAS KARACHI LAHORE DACCA
KUALA LUMPUR SINGAPORE HONG KONG TOKYO

ISBN 0 19 824366 9

Printed in Great Britain
at the University Press, Oxford
by Vivian Ridler
Printer to the University

Preface

THIS book is an inquiry into the nature of number, written largely from the logicist point of view first clearly expounded by Frege. It does not pretend to contain any discussion of the different theories which were traditionally recognized as rivals to logicism, namely formalism and intuitionism or other forms of constructivism. Some of the points that I advance in the course of my argument could certainly be turned against these opponents, and if my general conclusions are anywhere near the truth it would seem to follow that these opponents are in error. But I should admit that this is not in itself a very powerful riposte, for my argument relies upon assumptions which the opponents would not grant. For example I assume without argument that ordinary numerical formulae must surely have *some* meaning, that the classical two-valued logic applies to infinite totalities as well as to finite ones, that we are not at liberty to make up the rules of proof as we go along, and so on. About such fundamental assumptions of logicism this book has nothing to say.

Nor have I thought it necessary to devote any space to presenting a prima-facie case for logicism, in order to show that it deserves a detailed investigation, for I consider that this task was quite amply fulfilled by Frege himself. The first chapter of his *Foundations of Arithmetic* is wholly designed as an argument for the initial plausibility of the logicist thesis, largely on the ground that there are difficulties in all alternative views, and in my opinion this argument is still a persuasive one. Of course at this stage of the inquiry the logicist thesis itself is not entirely precise. As Frege himself puts it, it is the thesis that the truths of arithmetic are *analytic* truths, where we are to understand that a truth is analytic if and only if its proof rests only on general logical laws and on definitions. There are several points here that need further clarification—in particular it is not very clear what we may permit as a 'general logical law' or as a 'definition'—and one might also ask how exactly the thesis is supposed to help us in understanding the concept of a number. But it does certainly seem that if the thesis is indeed true (under some suitable interpretation) then it ought to be of *some* help, and even in this preliminary formulation the thesis is precise enough to be worth investigation. If we can actually carry out the programme of 'reducing' arithmetic to an ostensibly logical basis, then the questions that one might wish to raise about the significance of the reduction may be more fruitfully discussed after the reduction has been presented.

I begin, then, by demonstrating that the logicist programme, as Frege conceived it, can indeed be completed—*but* only if we abandon Frege's own conception of numbers as objects. This demonstration occupies the bulk of the present volume, which is intended as the first part of the work. Much of the argument here has to be devoted to issues which concern logic in general, irrespective of its relation to arithmetic, and in any case the whole emphasis of the discussion is on philosophical questions of analysis rather than on new and ingenious proofs. (Although the formal derivation I offer is not quite the same as any so far put forward—so far as I know—I do not wish to claim any great originality for it: all the essential ideas are to be found in Frege's own work.) Anyway, it is not until the last chapter of this part, when the logicist 'reduction' has been completed, that I come to discuss its significance for our understanding of the concept of number. Here my general conclusion is that there still remains a great deal more to be said, and accordingly in the second part of the work (now actively in progress) I attempt to say some of what is missing and to give a general characterization of all that is missing. I have subtitled the first part 'Natural Numbers' and the second part 'Rational and Irrational Numbers' because these titles best indicate how my discussion fits in with orthodox logicist treatments. But I should say at the outset that these titles seem to carry some implications which I eventually reject; in particular, I do not really regard myself as changing the subject between my two parts. The subtitles 'Numbers and Counting' and 'Numbers and Measuring' would be less open to objection on this score.

The book has been many years in gestation, and it has benefited much from the comments and suggestions of many friends with whom I have discussed its ideas. I hope I may be forgiven for here thanking them all anonymously. But I should like to acknowledge special debts to Michael Dummett and Robin Gandy, for they have each on different occasions given me searching criticisms of earlier versions, and these have led to some very substantial changes. My chief debt is almost certainly to Professor Quine, whose logical writings first brought me to think about the subject, and have proved a constant stimulus ever since. But of course the subject itself owes its existence to the pioneering work of Frege and Russell, and without them my inquiry would have been quite impossible.

DAVID BOSTOCK

Merton College, Oxford
August 1973

Contents

1. Introduction

1. Numbers as Objects

WHAT are the natural numbers? Well, in one sense the answer is of course obvious: the natural numbers are the numbers in the series 1, 2, 3, 4, 5, and so on—or if we prefer we may start the series with 0 rather than with 1. But clearly when a philosopher raises this question it is not this sort of answer he is looking for; he is looking, one might say, for an account or explanation of what the natural numbers are, and not merely for a list of them.

Shall we then give an account of the numbers by saying that they are what we use for numbering or counting groups of things? Groups of all sorts, we may go on, can be counted, and the procedure of counting consists in marking off the objects in the group one by one and assigning to each one number from the series of natural numbers (excluding 0). The numbers are assigned to objects in the order in which they occur in the series of natural numbers, and the number assigned to the last object in the group is said to be the number of the objects in that group, for by its position in the series of numbers it is able to serve as a measure of the size or 'numerosity' of the group. And we might go on to elaborate this account in various ways. But to this sort of answer it seems that the philosopher may reply with perfect justice that what he has been given is an account of the *use* that we make of the natural numbers—or at any rate of one of their uses, for he has been told about 'one, two, three, . . .' but not about 'first, second, third, . . .' or 'once, twice, thrice, . . .'—but he was not asking how we make use of the numbers but what the numbers *are*, and that does not seem to be quite the same thing. And it would seem that he could have replied in much the same way if we had taken a rather different line and spoken instead of the system of truths known as (elementary) arithmetic. We could for instance have said that all the necessary truths about numbers are consequences of the facts that 0 is a number which is not the successor of any number, that every number has one and only one successor, and that the numbers just are 0, the successor of 0, the successor of the successor of 0, and so on. And we could go on to furnish definitions of addition and subtraction, multiplication and division, exponentiation and the extraction of roots, and any other notions that seemed important. In this way we could put our philosopher in command of all the necessary truths about numbers, but still it seems that he can reply that he has not been told what numbers *are*. No doubt if he knew what 0 was, and if he knew what being the successor of a number was, then he would have an answer to his question, but these are points that he has not been told. So let us renew the question: what *are* the natural numbers?

Now the logicist's usual answer to this question is that numbers are to be taken as *classes* of a certain sort, and on the face of it this identification of numbers with classes seems unavoidable for him. For if truths about numbers are to reduce by definition to purely logical truths, then some 'logical objects' must presumably be used to give the definitions of the numbers themselves, and classes appear to be the only such objects available. Indeed Frege put the point in almost these words when faced with Russell's paradox: 'Even now I do not see how arithmetic can be scientifically established, how numbers can be *apprehended as logical objects* and brought under review, unless we are permitted, at least conditionally, to pass from a concept to its extension' (i.e. to introduce classes).[1] However, it has always been felt that the identification of numbers with classes is philosophically unacceptable. To put the point naïvely, we just do not think that numbers really are classes, and it is difficult to believe that this is mere blindness on our part. But there is also an argument, which has been developed at some length by Benacerraf,[2] which to my mind proves that numbers cannot be classes. The argument is simply that if numbers are classes then we should surely be able to say *which* classes they are, and yet this is in principle impossible.

Certainly one must admit that it is possible to set up a correlation between elementary arithmetic and some fragment of set

[1] *The Basic Laws of Arithmetic*, vol. ii, Appendix, p. 253 (p. 127 in Furth's translation).

[2] P. Benacerraf, 'What Numbers Could Not Be', *Philosophical Review*, January, 1965.

theory, by selecting a series of sets to correspond to the natural numbers and by defining relations and operations on these sets corresponding to the usual relations and operations on natural numbers, in such a way that this fragment of set theory may be shown to be isomorphic with elementary arithmetic. Thus every arithmetical truth will have its set-theoretical analogue, and furthermore the way in which numbers appear in ordinary discourse—for instance in statements of the form 'there are just n so-and-so's'—can always be explained in terms of arithmetical propositions and so in terms of their set-theoretical analogues. But the trouble is that this can be done in many ways. *Any* series of sets which satisfies Peano's postulates, i.e. any progression of sets, will do.[3] Whatever our progression, we can always call its first member '1', its second '2', and so on; and we can always explain a statement of the form "there are just n so-and-so's" as meaning "the set of all so-and-so's can be put in one-to-one correspondence with the set of all 'numbers' from '1' to 'n'." But if numbers *are* sets, then one and only one of these correlations will be the identity-correlation. In that case, *which* one?

Now if the non-arithmetical, ordinary language, statements involving numbers are treated in this way as dependent on the arithmetical language, and so explained in terms of whatever correlation is adopted, then it does seem clear that we could have no reason for preferring any one of the many possible correlations to any other. But it may be suggested that the ordinary non-arithmetical usage does after all furnish a clue here. For it is a marked feature of our use of numerals that although in arithmetical contexts they do seem to function as names in a broad sense, i.e. as words which denote objects (namely numbers), yet in non-arithmetical contexts they evidently do not function in this way at all. And at this point we may notice a very tempting analogy. It may well seem that there are many English words which function sometimes as names of classes but sometimes predicatively. For instance it may seem that 'man' functions roughly as the name of a class, or something similar, in 'man has evolved more rapidly than any other animal', but that it functions predicatively in 'Socrates is (a) man'. (And although we cannot say *quite* the same thing with e.g. 'horse', at least 'the horse' and 'a horse' exhibit the appropriate connection, and that

[3] Provided it is recursive, as Benacerraf emphasizes (op. cit.).

is perhaps good enough.) But now, if these two uses are appropriately described as we have described them, then the relationship between them is clear: 'man' as a class-name names the class of all and only those objects to which, as a predicate, it truly applies. Pursuing the analogy, then, we suggest that numerals also function sometimes as names of classes and sometimes predicatively, and a numeral used as a class-name will name the class of all and only those objects to which, when used as a predicate, it truly applies. By no stretch of the imagination can we construe numerals as predicates applying to individuals, but they can perhaps be reckoned as predicates applying to classes. So the result is that when the numeral 'n' occurs as a name it names the class of all and only those classes to which, as a predicate, it truly applies; that is, it names the class of all n-membered classes. And this, of course, is the identification originally put forward by Frege and Russell, which does intuitively seem the least artificial of all those so far proposed (but also the most troublesome to modern set theory).

But it can easily be seen that the proposed analogy will not work. For one thing, one may well doubt whether the words 'man' and 'the horse' are indeed replaceable in these contexts by the phrases 'the class of all men' and 'the class of all horses'; certainly the replacement looks somewhat strange in the example we have chosen. But more importantly it is crucial to this analogy to maintain that numerals, when they do not occur as names, occur as predicates—in fact as predicates asserting membership in a class. And this thesis is plainly false. There are many permissible ways of saying e.g. that Jupiter has four moons, but 'the class of Jupiter's moons is a four' is not among them—at least, not in English nor in any other natural language I happen to know. We can say 'Jupiter's moons are four (in number)', but here the plural verb is significant, for it shows that we are not after all attaching a predicate to a single class; and in fact when numerals do occur apparently in a predicative position they take irreducibly plural subjects in a quite peculiar manner. But more important than this is the fact that numerals very seldom do occur in positions accessible to predicates; in most standard idioms they behave just like applicatives such as 'a', 'the', 'all', 'any', 'many', 'most', 'few', 'some', etc.—i.e. they behave as quantifier-expressions. Grammatical studies yield ample evidence for this,[4] and I shall not dwell upon it here, for indeed once made the point seems obvious. Its importance in the present connection is simply that it destroys the alleged analogy between numerals and class-words such as 'man', and so destroys this argument—which is the only argument I know of—in favour of the Frege–Russell correlation of numbers and classes. If in non-arithmetical contexts numerals function as expressions for quantifiers, then this clearly furnishes us with no evidence to determine what *classes* they name in arithmetical contexts, for I take it that no one will say that the word 'some' sometimes names the class of some-membered classes or that the word 'most' sometimes names the class of most-membered classes.

It is in a way remarkable that the fact that numerals function not as names but as expressions for quantifiers was most clearly recognized by Frege himself and used by him to discredit many rival analyses of number. Indeed chapters 2 and 3 of his *Foundations of Arithmetic* are wholly designed as an argument to just this conclusion, for this is what Frege means by his dictum that 'the content of a statement of number is an assertion about a concept'. As he at once explains

> This is perhaps clearest with the number 0. If I say 'Venus has 0 moons', there simply does not exist any moon or agglomeration of moons for anything to be asserted of; but what happens is that a property is assigned to the *concept* 'moon of Venus', namely that of including nothing under it. If I say 'the King's carriage is drawn by four horses' then I assign the number four to the concept 'horse that draws the King's carriage' (p. 59).

And a little later he continues

> By properties which are asserted of a concept I naturally do not mean the characteristics which make up the concept. These latter are properties of the things which fall under the concept, not of the concept. Thus 'rectangular' is not a property of the concept 'rectangular triangle'; but the proposition that there exists no rectangular equilateral rectilinear triangle does state a property of the concept 'rectangular equilateral rectilinear triangle'; it assigns to it the number nought. In this respect existence is analogous to number. Affirmation of existence is in fact nothing but denial of the number nought (pp. 64–5).

Thus Frege's view clearly is that a 'statement of number' is

[4] See Benacerraf, op. cit., pp. 59–61.

logically of the same type as a statement of existence, in that both assert something of a concept, and this simply is the view that 'statements of number' are statements in which quantifiers, or *second-level concepts*, are applied to (first-level) concepts; for quantifiers are the prime example of second-level concepts for Frege. But unfortunately Frege went on to argue that it is not exactly the *number* which is asserted of the concept; rather, the number is only an element in what is asserted of the concept, for it is itself not a quantifier but an *object* (p. 68). However, it may be said at once that although Frege did have some reasons for supposing that numbers must be recognized to be objects (and I shall be discussing these later), he gave no reason at all for supposing that they must be the particular objects he selected. In fact he actually remarked on this point

> This . . . cannot be expected to meet with universal approval, and many will prefer other methods. . . . I attach no decisive importance even to bringing in the extensions of concepts at all (p. 117).[5]

It seems to me that Frege's hesitation here is correct: there *is* no reason to bring in the extensions of concepts (i.e. classes), and Frege has only been led into error by so doing.

Russell's position on this issue is somewhat more difficult to evaluate. In his early work *The Principles of Mathematics* he begins in sections 70–4 by construing numbers not exactly as quantifiers but as predicates of classes, with the rider that they are predicates of 'classes as many' and not of 'classes as one'. It would seem that Russell was partly led to his conception of a 'class as many' by a search for something that numbers could be predicated of, for one does not say that the *class* of the men in the room *is* three, but rather that the *men* in the room *are* three (or, according to Russell, that Smith and Brown and Robinson *are* three). However, he emphasized that a 'class as many' could not properly count as a logical subject, and therefore it is not absolutely clear whether we should say that he does regard numbers here as predicates. (At any rate we could not say that he regards them as quantifiers, for at this time he was still enmeshed in the toils of his first intricate theory of denoting, which prevented him from reaching any clear conception of a quantifier altogether.) Anyway, when later in section 111 he comes to

[5] But by the time of *The Basic Laws of Arithmetic* his views on this had evidently changed. See text to note 1.

introduce his formal definition of numbers as classes he immediately comments that 'it must appear, at first sight, a wholly indefensible paradox', proceeds to reject the only way of taking it that would reduce the paradox, and concludes

> Mathematically, a number is nothing but a class of similar classes: this definition allows the deduction of all the usual properties of numbers, whether finite or infinite, and is the only one (so far as I know) which is possible in terms of the fundamental concepts of general logic. . . . Wherever Mathematics derives a common property from a reflexive, symmetrical, and transitive relation [such as similarity], all mathematical purposes of the supposed common property are completely served when it is replaced by the class of terms having the given relation to a given term; and this is precisely the case presented by cardinal numbers (p. 116).

Ultimately, then, Russell's defence of his definition is just that it serves all mathematical purposes, and we have seen that this defence fails. For the same could be said with equal justice of infinitely many other definitions. In fact in this passage Russell is very close to the position of those more recent philosophers who speak of 'replacement', holding that so long as the definition is adequate to all mathematical uses it really does not matter whether it is numbers that are defined or something else. And this position is one that he continued to adopt in later years, for instance in his *Introduction to Mathematical Philosophy*, where he remarks

> It is more prudent to content ourselves with the class of couples, which we are sure of, than to hunt for a problematical number 2 which must always remain elusive (p. 18).

But the objection to this position is quite clear: if it is not *numbers* that we are talking about, then why should we suppose that what we have deduced is *arithmetic*? Yet it was precisely arithmetic that was claimed to be reducible to logic.

However, we should notice that after *The Principles of Mathematics* Russell's definition of number was in fact altered as a consequence of his changed view of classes generally, by which they are no longer taken to be 'logical objects' but rather 'logical fictions' which disappear upon analysis. This makes it more difficult to assess his liability to the objection developed by Benacerraf, for Benacerraf clearly conceived his objection as an objection to supposing that numbers were such *objects* as classes are supposed to be, generalizing it to the conclusion that numbers cannot be *objects* at all. Let us therefore leave the later Russell on one side and return to Benacerraf's objection.

As I have just said, Benacerraf ends his paper by arguing that the considerations which show that numbers cannot actually be classes will also show that they cannot be objects at all. For the fact is that any objects whatever which form a recursive progression can *play the role of* the numbers, and there is no conceivable ground for singling out any one progression from all the others as *the progression of the numbers*. So he concludes that 'arithmetic is the science that elaborates the abstract structure that all progressions have in common merely in virtue of being progressions', and consequently that it may perhaps be viewed as the elaboration of the properties of *all* systems of objects exhibiting that abstract structure, but not as concerned with just one such system, the supposed system of objects called numbers. But I think that we could put this point in a slightly different way, as follows. What is needed for a set of objects to play the role of the numbers is that we should be able to *use* them as numbers (for instance in counting), and this means, I think, that we must have set up an effective correlation between these objects and the numerical quantifiers that occur in ordinary discourse in expressions of the form 'there are n so-and-so's which . . .'. It will then follow, of course, that the objects are arranged in a recursive progression, and we may perhaps take the properties that they have *in virtue of* this correlation to be the properties of 'the numbers', viewing arithmetic as the science dealing with the properties of all objects thus correlated with numerical quantifiers. But evidently the properties which the objects have in virtue of their correlation will stem from corresponding properties of the numerical quantifiers themselves, and so it would certainly seem that arithmetic is better viewed not as a science dealing with all systems of objects that may be correlated with numerical quantifiers but simply as the science of the numerical quantifiers themselves. And it further seems that we *must* make some move of this sort if the central logicist thesis is to be upheld; for it would not appear to be a matter of logic that *there are* any objects correlated with the numerical quantifiers at all, but it surely is an arithmetical truth that there are infinitely many numbers. So I conclude that it is both a plausible consequence to draw from

Benacerraf's argument, and a requirement for the logicist thesis, that arithmetic should *not* be regarded as a science treating of objects at all.

The meaning of the claim that arithmetic does not treat of objects, and hence that numbers are not objects, is perhaps none too clear, and I shall be saying more about it in the next chapter. For the present it is perhaps sufficient to say that, as I understand it, to abandon the claim that numbers are objects is also to abandon the claim that numerals ever do function as names, i.e. as words which denote or refer. The *apparent* referring role of numerals in arithmetical propositions must somehow be explained away so that numbers become, in Russell's sense, logical fictions created by definitional abbreviation. More precisely, we must explain the sense of such arithmetical propositions as '3 is a prime number' or '3 is greater than 2' by analysing these propositions *as wholes* into other propositions in which there is no longer any apparent reference to a number, but instead we make use of the associated quantifiers 'there are 3 . . .' and 'there are 2 . . .'. Further, our explanations must obviously be such that the truths of arithmetic can be seen to follow from the truths concerning numerical quantifiers in virtue of our definitions, for otherwise we would hardly have established the claim that any introduction of objects is quite superfluous. So our task is first to construct numerical quantifiers and then to proceed from this basis to a reductive account of the apparent referring role of numerals and indeed of arithmetic generally.

2. Numbers as Fictions

At a first glance there is no difficulty over the initial notion of a numerical quantifier. Given first-order predicate calculus with identity it is a simple matter to introduce quantifiers corresponding to 'There are at least n Fs' and 'There are exactly n Fs' for any specific numeral 'n'. Thus for the first we produce a string of definitions constructed on the pattern:

'$(\exists 1x)(Fx)$' for '$(\exists x)(Fx)$'
'$(\exists 2x)(Fx)$' for '$(\exists x)(Fx \,\&\, (\exists 1y)(Fy \,\&\, y \neq x))$,
'$(\exists 3x)(Fx)$' for '$(\exists x)(Fx \,\&\, (\exists 2y)(Fy \,\&\, y \neq x))$'
'$(\exists 4x)(Fx)$' for '$(\exists x)(Fx \,\&\, (\exists 3y)(Fy \,\&\, y \neq x))$'
etc.

And for the second we produce a similar string:

$$
\begin{array}{lll}
`(0x)(Fx)\text{'} & \text{for} & `\sim(\exists x)(Fx)\text{'} \\
`(1x)(Fx)\text{'} & \text{for} & `(\exists x)(Fx \ \& \ (0y)(Fy \ \& \ y \neq x))\text{'} \\
`(2x)(Fx)\text{'} & \text{for} & `(\exists x)(Fx \ \& \ (1y)(Fy \ \& \ y \neq x))\text{'} \\
`(3x)(Fx)\text{'} & \text{for} & `(\exists x)((Fx \ \& \ (2y)(Fy \ \& \ y \neq x))\text{'}
\end{array}
$$

etc.

I shall call the first of these series the series of *weak* numerical quantifiers, and the second the series of *strong* numerical quantifiers. It is intuitively clear that, where 'n′' is the next numeral after 'n', the relation between them is given by

$$(nx)(Fx) \equiv ((\exists nx)(Fx) \ \& \ \sim(\exists n'x)(Fx))$$

except for the case in which 'n' is '0', for then the right-hand side of this equivalence is undefined.[6]

Now it is commonly said that these numerical quantifiers cannot be used as a basis for arithmetic, since the definitions by which they are introduced give us no warrant for introducing variables in place of our numerals, and this will clearly be essential. And there are also some, for instance Quine,[7] who hold that if we do introduce quantifiable variables here then we are already tacitly construing numbers as objects however much we may deny this, for only quantification over objects makes sense. I certainly agree that the introduction of numerical variables is essential, and I also agree that such an introduction is not fully justified by the mere fact that we do actually understand how to continue the strings of definitions just indicated; in fact the justification of this step is one of our major problems, and I shall not complete my discussion of it until chapter 5. However, the Quinean objection that we can quantify only over objects is one that I do not accept, and I shall devote chapter 3 to rebutting it. Assuming, then, that these difficulties can be overcome, let us now proceed to use 'n', 'm', 'p',..., as schematic letters for

[6] To remove the exception we might adopt the artificial definitions

$$
\begin{array}{lll}
`(\exists 0x)(Fx)\text{'} & \text{for} & `(\forall x)(Fx \vee \sim Fx)\text{'} \\
`(\exists 1x)(Fx)\text{'} & \text{for} & `(\exists x)(Fx \ \& \ (\exists 0y)(Fy \ \& \ y \neq x))\text{'}
\end{array}
$$

but I shall in fact retain '(∃1x)' as the first weak quantifier.

[7] See *From a Logical Point of View*, chapters 1 and 6; *Methods of Logic*, section 38.

numerals with '𝔫', '𝔪', '𝔭',..., as associated variables, in order to see what can be done with them. Certainly there seems to be no intuitive difficulty in this procedure, nor in the further assumption—which I shall at once make—that all ordinary laws of quantification theory apply unchanged to our new variables.

I shall take it that we have provided a satisfactory basis for the construction of (elementary) arithmetic when we have succeeded in obtaining Peano's postulates, which may be put in the form:

1. 1 (or 0) is a number
2a. Every number has at least one successor which is a number
2b. Every number has at most one successor which is a number
3. No two numbers have the same successor which is a number
4. 1 (or 0) is not the successor of any number
5. Whatever is true of 1 (or 0), and is also true of the successor of any number whenever it is true of that number, is further true of all numbers.

It will be seen that these postulates employ four concepts that go beyond the straightforward logic of quantification, namely the concept of number, of the first number (1 or 0), of being the same number, and of being a successor of a number. Our first task is evidently to provide definitions of these notions.

However, the notion of number itself would seem to be one that we cannot reasonably give a definition of, since the task of singling out numbers from everything else is in our case already being performed by the use of a special style of variable, so no further definition is called for. As far as the first number is concerned, it seems most natural to take 0 as our first number if we are working with the strong numerical quantifiers and 1 as our first number if we are working with the weak ones. As a matter of fact it turns out that the deductions are altogether simpler if we choose the weak quantifiers to work with, so I shall pursue this course for the purposes of the present illustration, and accordingly take 1 as the first natural number. But since our aim is to construct arithmetic without taking numbers as objects, we shall again not attempt to furnish any further definition of 1 besides that already given in the series of weak numerical quantifiers. Let us, therefore, formally set down the principle of that series in the two definitions.

D1: '$(\exists 1x)(Fx)$' for '$(\exists x)(Fx)$'

D2: '$(\exists n'x)(Fx)$' for '$(\exists x)(Fx \;\&\; (\exists ny)(Fy \;\&\; y \neq x))$'

Turning now to being the same number and being a successor of a number, I propose to set up tentatively the two definitions

D3: '$n=m$' for '$(\forall \mathfrak{f})((\exists nx)(\mathfrak{f}x) \equiv (\exists mx)(\mathfrak{f}x))$'

D4: 'nSm' for '$(\forall \mathfrak{f})((\exists nx)(\mathfrak{f}x) \equiv (\exists m'x)(\mathfrak{f}x))$'

It may be remarked that these two definitions make use of quantification over predicates, so extending our basis from first-order to second-order quantification theory and exposing the construction once more to Quine's strictures on the proper employment of quantification. *This* problem, however, is one that we have met already, and we may ignore it here, for there are certainly other more serious problems connected with these definitions. I shall leave these to emerge shortly, but let us just note here one satisfactory feature of the definitions, namely that the identity that is here defined is a reflexive, symmetrical, and transitive relation, and together with the definition of succession it at once yields the substitution-theses

$$\vdash n=m \supset (nSp \equiv mSp)$$

$$\vdash n=m \supset (pSn \equiv pSm).$$

It follows that the identity here defined licences substitution in all contexts definable in terms of identity and succession, and therefore that it licenses substitution in all arithmetical contexts, which is evidently just as it should be.

Supposing, then, that we now have satisfactory definitions of all the notions employed in Peano's postulates, we may pass on to the problem of establishing these postulates. The first postulates to be considered here are that 1 is a number and that every number has a successor which is a number. Since we can indicate the notion of number only by employing our numerical variables, these postulates must be set down in some such way as

$$\text{Ax1:} \begin{cases} \vdash (\exists \mathfrak{n})(\mathfrak{n}=1) \\ \vdash (\exists \mathfrak{n})(\mathfrak{n}Sm) \end{cases}$$

and evidently we must for the moment adopt them as axioms for we have nothing from which to deduce them.

Turning now to the last postulate, the principle of mathematical induction, we find that it too will require the introduction of an axiom. This is because the usual logicist constructions establish the principle of induction by using the notion of the ancestral in the *definition* of natural number, but we have no such definition. It is convenient to distinguish two versions of the ancestral of a relation: the *proper* ancestral of a relation '... R ...' is the relation '... ⋆R ...' such that

$$a{\star}Rb \equiv aRb \vee (\exists x)(aRx \;\&\; xRb) \vee (\exists x)(\exists y)(aRx \;\&\; xRy \;\&\; yRb) \vee \ldots$$

or more briefly

$$\star R = (R \vee R|R \vee R|R|R \vee R|R|R|R \vee \ldots)$$

while the (improper) ancestral of a relation '... R ...' is the relation '... *R ...' such that

$$a{*}Rb \equiv (a{=}b \vee a{\star}Rb).$$

The notion of the ancestral may be introduced into first-order quantification theory by appropriate axioms, or it may be introduced into second-order quantification theory by explicit definition, and for present purposes I assume that this notion is available to us, so that in particular we can apply it to the successor-relation to form the proper and improper ancestrals '... ⋆S ...' and '... *S ...'. Now from the properties of the ancestral of a relation we may easily deduce the thesis

$$\vdash (Fa \;\&\; (\forall x)(\forall y)(xRy \supset (Fy \supset Fx))) \supset (\forall x)(x{*}Ra \supset Fx)$$

and I shall assume that this thesis will still hold when '... R ...' is taken as the successor relation, the variables as numerical variables, and 'a' as 1. (The ground for the assumption would be that the thesis depends only on the definition of the ancestral and ordinary rules of quantification theory, and we have already assumed that the latter apply to numerical variables. I shall investigate the assumption in detail in chapter 5.) So taken, the thesis states that whatever is true of 1, and is also true of the successor of any number whenever it is true of that number itself, is further true of every number *which bears the ancestral of the successor relation to 1.* To obtain the principle of mathematical induction in its usual form, then, we need only show that every natural number does indeed bear the ancestral of the successor relation to 1. In the usual logicist constructions, where the natural numbers are identified with certain classes, this result is secured by *defining* the natural numbers as those classes that bear the ancestral of the successor relation to 1; but as I pointed out earlier this course is not open to us, and so we must again employ an axiom

$$\text{Ax } 2 \vdash n{*}\mathbf{S}\,1.$$

I think it is fairly evident to intuition that what our two axioms do between them is to fix the range of the new numerical variables. For the first axiom states a sufficient condition for being in this range, namely being either 1 or the successor of something in the range, and the second axiom states a necessary condition for being in the range, namely being either 1 or the successor of 1 or the successor of the successor of 1 or . . . and so on. Furthermore the two axioms between them have very much the same effect as the usual logicist definition. The problem of how to put this point more precisely and so remove the axiomatic status of these axioms is one that I shall discuss in chapter 5, but for the moment let us simply accept them as axioms and observe whether they are sufficient for the deduction of the remaining postulates.

Now postulate 2b, which we may write as

$$\vdash (n\mathbf{S}p \ \& \ m\mathbf{S}p) \supset n = m$$

is an immediate consequence of our definitions of identity and succession, while postulate 4, which we may write as

$$\vdash \sim(1\mathbf{S}n)$$

is again a straightforward consequence of our definition of succession and our second axiom.[8] But postulate 3, that no two numbers have the same successor, is by no means so simple a matter.

First we may notice that this postulate is, with our definitions, equivalent to the postulate that no number succeeds itself. For without any dubious manœuvring it is possible to establish the theorems[9]

$$\vdash n{*}\mathbf{S}m \lor n = m \lor m{*}\mathbf{S}n$$

$$\vdash (n{*}\mathbf{S}m \ \& \ p\mathbf{S}m) \supset (n = p \lor n{*}\mathbf{S}p)$$

$$\vdash (n{*}\mathbf{S}p \ \& \ p\mathbf{S}n) \supset n = p$$

[8] Proofs may be found in the appendix to this chapter. The theorem just mentioned is listed there as theorem 3.

[9] Theorems 5, 4, and (an immediate consequence of) 6.

and hence if we suppose that there are numbers n, m, p, such that

$$pSn \ \& \ pSm \ \& \ n \neq m$$

then by applying the first of these theorems we deduce

$$(pSn \ \& \ n^{*}Sm \ \& \ pSm) \vee (pSm \ \& \ m^{*}Sn \ \& \ pSn)$$

and by applying the second we deduce from this

$$(pSn \ \& \ n=p) \vee (pSn \ \& \ n^{*}Sp) \vee (pSm \ \& \ m=p) \vee (pSm \ \& \ m^{*}Sp)$$

and by applying the third we deduce further from this

$$(pSn \ \& \ n=p) \vee (pSm \ \& \ m=p)$$

whence finally $$pSp.$$

(And the converse deduction may also be achieved by an inductive argument making use of postulate 4.)

Considering, then, the alternative postulate that no number succeeds itself we find that this would expand by our definition to

$$\vdash (\exists \mathfrak{f})(((\exists nx)(\mathfrak{f}x) \ \& \ \sim(\exists n'x)(\mathfrak{f}x)) \vee ((\exists n'x)(\mathfrak{f}x) \ \& \ \sim(\exists nx)(\mathfrak{f}x))).$$

But the second disjunct here may easily be shown to be contradictory,[10] and so the postulate may again be simplified to

$$\vdash (\exists \mathfrak{f})((\exists nx)(\mathfrak{f}x) \ \& \ \sim(\exists n'x)(\mathfrak{f}x))$$

or, introducing the strong numerical quantifiers by the definition suggested on p. 10,

$$\vdash (\exists \mathfrak{f})(nx)(\mathfrak{f}x).$$

It would indeed be possible to simplify the postulate yet further, for from our present basis we could establish the theorem[11]

$$\vdash (\forall \mathfrak{f})((\exists nx)(\mathfrak{f}x) \supset (\exists \mathfrak{g})((\forall x)(\mathfrak{g}x \supset \mathfrak{f}x) \ \& \ (nx)(\mathfrak{g}x)))$$

and this shows that the apparently weaker version

$$\vdash (\exists \mathfrak{f})(\exists nx)(\mathfrak{f}x)$$

is equivalent to our last. But this final step of simplification is not necessary for my purpose.

It turns out, then, that the missing postulate is equivalent to the postulate that for any natural number n there is some predicate F such that there are exactly (or at least) n Fs; in other words it is a version of the axiom of infinity. Further, it is easily seen that unless this postulate is true the definitions of identity and

[10] Notice that '$n' \geqslant n$' is an immediate consequence of theorem 6.

[11] I omit the proof, which is a trifle long.

succession we laid down to begin with are quite unacceptable. For if there were only a finite number, say n, of objects, then all propositions of the forms '$(\exists n'x)(Fx)$', '$(\exists n''x)(Fx)$', '$(\exists n'''x)(Fx)$', and so on, would be false, and hence equivalent; and therefore n', n'', n''', and so on, would by our definition all be the same number. So, if we are to stick to our definitions, it would seem to be essential that we ensure an adequate supply of objects.

One way of doing this is of course to follow Russell's procedure and lay down the axiom of infinity as an axiom, but it seems clear that this would not be in the spirit of logicism; indeed if we construe this axiom as stating that there are infinitely many objects *of lowest type* (which we must do if 'x' is a variable for objects of lowest type), then it is not absolutely clear that there is sufficient ground even for supposing the axiom *true*.[12] On the other hand we would have a conclusive reason for supposing the axiom true if we followed Frege's course and took *numbers* to be objects, for there certainly are infinitely many numbers. In fact Frege, in his construction of arithmetic, gave a proof of the theorem that for any natural number n there are exactly n numbers in the series of natural numbers from 1 to n (inclusive), i.e. in our symbolism he gave a proof that

$$\vdash (n\mathfrak{m})(n * S\mathfrak{m})$$

But from this theorem, which I refer to as 'Frege's theorem', the desired axiom of infinity will follow by existential generalization *provided* we grant the assumption that numbers are objects. However, we have already argued that this course is not available to us, and nor apparently is Russell's course, so there seems to be nothing for it but to modify our definitions of identity and succession.

One modification which one might suggest here is the addition of a modal operator, so that n and m are to be the same number only if it is *necessary* that the one applies wherever the other does; in other words, our revised definitions are to be

$$\text{'}n = m\text{' for '}\Box(\forall\mathfrak{f})((\exists nx)(\mathfrak{f}x) \equiv (\exists mx)(\mathfrak{f}x))\text{'}$$

$$\text{'}nSm\text{' for '}\Box(\forall\mathfrak{f})((\exists nx)(\mathfrak{f}x) \equiv (\exists m'x)(\mathfrak{f}x))\text{'}$$

[12] This point will emerge more clearly from the discussion of objects in chapter 2. Meanwhile compare Russell, *Introduction to Mathematical Philosophy*, chapter 13.

We then find that all the previous arguments go through with this modification, and in place of a Russellian axiom of infinity we now require the modalized axiom

$$\vdash \Diamond (\exists \mathfrak{f})(nx)(\mathfrak{f}x).$$

At least this axiom has the advantage of being certainly true, since it is, for example, quite obviously true that for any natural number it is *possible* that there are just that number of apples in the world. Further, one might perhaps support this proposal on the ground that it would make every proposition of arithmetic a modal proposition, which might appear to accord very nicely with our view that the propositions of arithmetic are not just accidentally true. It seems to me doubtful that we could justify the new axiom from a strictly logicist point of view, but that, it might be urged, just shows that logicism is after all indefensible.

However, I think there is a better course open to us, for the chief objection to such a modalized construction of arithmetic is not so much that it is incorrect as that it is incomplete, and once the incompleteness is remedied the modal element can be seen to be superfluous. For the modal system just considered endorses the view that numbers are not objects and yet it is also a system in which only objects can be counted, in so far as its only numerical quantifiers are defined by means of variables ranging over objects and nothing else. But this must surely be a mistake, for Frege was obviously right to claim that we *can* count numbers. Indeed what I have called Frege's theorem is a very fundamental theorem in the theory of counting. In counting we pair off the objects to be counted with natural numbers in their natural order, so that if the last object counted is paired off with the number n then the whole group of objects counted is paired off with the series of natural numbers from 1 to n. The pairing-off operation, therefore, establishes that there are as many objects in the group counted as there are numbers from 1 to n, and it is just *because* there are exactly n numbers from 1 to n that we are entitled to conclude that there are exactly n objects in the group counted. This theorem of Frege's, then, is of fundamental importance, and it would seem reasonable to make it a condition on any adequate theory of number that it should yield this theorem. Frege, as I have said, obtained the theorem by construing numbers as objects, but there is surely an alternative procedure, and that is to remove the restriction that only objects may be counted. Indeed, we have seen that the initial series of numerical quantifiers can be defined simply with the aid of universal quantification and identity, and it follows that numerical quantifiers will be available wherever these are available. We have already assumed that quantification is available over numbers, and we have already suggested a definition of identity for numbers, so this precondition is apparently satisfied. We should then, be able to press on without more ado, applying numerical quantifiers to the numbers we have just introduced, and so establishing Frege's theorem.

But of course the original definitions of identity and succession between numbers are no more satisfactory for this purpose than they were originally. In fact it turns out that at a crucial point in the argument we need to make use of the thesis

$$\vdash n=m \supset ((\exists n\mathfrak{p})(\text{——}\mathfrak{p}\text{——}) \equiv (\exists m\mathfrak{p})(\text{——}\mathfrak{p}\text{——}))$$

and with our original definition of identity the truth of this thesis will depend upon there being infinitely many objects, which is evidently not acceptable. What does emerge, however, is that the new definition of identity that we require must be one that yields this thesis, and so the most obvious definition to adopt will be precisely

$$\text{`}n=m\text{'} \quad \text{for} \quad \text{`}(\forall\phi)((\exists n\mathfrak{p})(\phi\mathfrak{p}) \equiv (\exists m\mathfrak{p})(\phi\mathfrak{p}))\text{'}$$

(where 'ϕ' is a variable for properties of numbers). If we do adopt this definition, and modify the other definitions analogously, then all our earlier theorems remain available and in addition it now becomes possible to complete the proof of Frege's theorem and thereby obtain the missing postulate.[13]

This version of the deduction of Peano's postulates is essentially similar to that put forward by Borkowski,[14] but I think that

[13] See theorems 7–9 of the appendix. The crucial thesis is used in the proof of theorem 7 at line 7, and in the proof of theorem 9 at line 3.

[14] L. Borkowski, 'Reduction of Arithmetic to Logic . . .', *Studia Logica*, VIII, 1958. Borkowski's deduction is more long-winded than mine, partly because it treats of cardinal numbers generally, but mainly because it works with strong numerical quantifiers rather than weak ones. Borkowski in fact does not define identity between numbers, but rather takes identity as undefined and lays down an axiom which implies (in my symbolism)

$$\vdash(\forall\phi)\,((n\mathfrak{p})(\phi\mathfrak{p}) \equiv (m\mathfrak{p})(\phi\mathfrak{p})) \supset n=m$$

This has the same effect as the definition suggested, for Borkowski assumes the one is liable to find it somewhat unsatisfactory. The most obvious ground for complaint is that we are now confining our attention entirely to numerical quantifiers which apply to numbers (i.e. to themselves), and in the definition of these quantifiers we are accordingly employing the 'identity' between numbers just defined. But since our position is that numbers should not be taken to be objects, and since numerical quantifiers applying to objects would presumably be defined with the help of a different definition of identity, it is not clear that the results we obtain for the one set of quantifiers will have any bearing on the other set. And consequently it is not clear that we have any title to claim that our results are indeed results concerning *numbers* as we ordinarily understand them. It would in fact be perfectly possible to give explicit introductions of further numerical quantifiers applying to items of other types by making use of the theorem of Frege's that we have now established—we might for instance set down the definition

$$\text{`(nx)(Fx)'} \quad \text{for} \quad \text{`}\hat{x}(Fx)\,\text{sim}\,\hat{\mathfrak{m}}(n{*}S\mathfrak{m})\text{'}$$

(where 'sim' represents 'as many as' and may be defined in terms of one-to-one correlations in the usual way)—but this procedure would only serve to intensify our doubts about the suggested analysis of numbers. For if the expression 'there are 3 . . .' is given one analysis in the context 'there are 3 *numbers* which . . .' and a different analysis in the context 'there are 3 *objects* which . . .', we are not likely to be sympathetic to the claim that the first of these is prior to the second, and that it is indeed the first which gives the sense of arithmetical propositions apparently about the number 3. Besides, I think that it is in fact quite clear that we are *not* employing different numerical quantifiers when we say, for instance, that *there are 3* men in the room, or that *there are 3* numbers from 1 to 3 inclusive, or indeed

converse as a property of identity and uses no other properties of identity. Borkowski's other definitions depart from my suggestions only in that he does not, as one might expect, define his first number, 0, by putting

$$\text{`}(0\mathfrak{p})(\phi\mathfrak{p})\text{'} \quad \text{for} \quad \text{`}{\sim}(\exists\mathfrak{p})(\phi\mathfrak{p})\text{'}$$

but in this case, and in this case only, employs typically ambiguous symbols in place of 'ϕ' and '$\mathfrak{p}$'. However, he makes no use of this added generality, and it is difficult to see why he includes it here for—as Michael Dummett has pointed out to me—he provides no sound means of establishing similarly general equivalences for further numerical quantifiers.

that in a universe of 2 objects *there are* 3 non-equivalent and non-universal predicates. And if this is correct we must somehow contrive to introduce our numerical quantifiers in such a way that they can be seen to be applicable unchanged to items of any type, including themselves.

This leads to a further difficulty, for it is notorious that Russell, for one, was led by his theory of types to deny the possibility of this, urging that we must recognize *different* numbers 3 applicable to each different type. And Frege too would have denied it, but on the quite different ground that numerical quantifiers are after all only applicable to objects (even if those objects are referred to as 'quantifiers' or as 'predicates'). This is not because he held, with Quine, that it is only *quantification* over objects that makes sense—indeed he certainly did not hold this—but rather because he held that numerical quantifiers must depend upon identity and that *identity* is strictly speaking indefinable and makes sense only between objects.[15] Thus although he is quite prepared to use the symbol '$(\exists \mathfrak{f})(\text{—} \mathfrak{f} \text{—})$', he denies any use for the symbol '$(\exists 3\mathfrak{f})(\text{—} \mathfrak{f} \text{—})$', and would similarly deny any use for the symbol '$(\exists 3\mathfrak{p})(\text{—} \mathfrak{p} \text{—})$' if it is insisted that '$\mathfrak{p}$' is a variable for a number and numbers are not objects.

Well, I think this is a sufficient indication of the problems we have to face, and of the route that we must take to overcome them. First, something must be done to clarify the question whether numbers are objects by discussing the relevant notion of an object, and this I undertake in the following chapter. Second, we must clear away the Quinean objection that only quantification over objects makes sense, and I turn to this next in chapter 3. Then begins the more positive part of the discussion, in which the main object is to give a precise introduction of numerical quantifiers, and here there are several desiderata. First (but perhaps least importantly) we should aim to show that our two axioms can indeed be seen as consequences of the definition of numerical quantifiers, and this is most easily done by viewing the notion of a numerical quantifier as a special case of the more general notion of a quantifier. Since we hold that the same numerical quantifier is applicable in many different types we must first reach an

[15] See Furth's discussion of this point in his introduction to *The Basic Laws of Arithmetic*, pp. xxxviii–xliv.

understanding of the idea of a logical type—thus continuing the discussion of objects in chapter 2—and show how quantifiers in general may be viewed in a type-neutral way; this discussion, which also builds upon the results of chapter 3, occupies chapter 4. Turning then to numerical quantifiers in particular, we must single these out from other quantifiers, making sure that they too turn out to be type-neutral despite their connection with identity, and finally we must apply our results to the analysis of arithmetical propositions and the construction of arithmetic. This occupies chapter 5, and at the last, when we have achieved a construction of arithmetic which apparently satisfies the logicist requirements, we may consider the question what exactly this shows us about natural numbers and arithmetic.

Appendix: The Deduction of Peano's Postulates

In the following proofs I make use of the definitions and axioms:

D1: '$(\exists 1x)(Fx)$' for '$(\exists x)(Fx)$'

D2: '$(\exists n'x)(Fx)$' for '$(\exists x)(Fx \,\&\, (\exists ny)(Fy \,\&\, y \neq x))$'

D3: '$n = m$' for '$(\forall \mathfrak{f})((\exists nx)(\mathfrak{f}x) \equiv (\exists mx)(\mathfrak{f}x))$'

D4: 'nSm' for '$(\forall \mathfrak{f})((\exists nx)(\mathfrak{f}x) \equiv (\exists m'x)(\mathfrak{f}x))$'

D5: '$n \geqslant m$' for '$(\forall \mathfrak{f})((\exists nx)(\mathfrak{f}x) \supset (\exists mx)(\mathfrak{f}x))$'

Ax 1: $(\exists \mathfrak{n})(\mathfrak{n} = 1) \,\&\, (\exists \mathfrak{n})(\mathfrak{n}Sm)$

Ax 2: $n{*}S1$

It may be noted that although I do not start with the assumption that there is just one number n' for any number n, nevertheless the symbol 'n'' is defined for any context for which 'n' is defined, so for instance the first axiom could equally well have been written as

$$(\exists \mathfrak{n})(\mathfrak{n} = 1) \,\&\, (\exists \mathfrak{n})(\mathfrak{n} = m').$$

I do not cite this axiom explicitly in the proofs that follow, but I use it tacitly in assuming the rules of inference

If '$(\forall \mathfrak{n})(\text{—}\,\mathfrak{n}\,\text{—})$' is a theorem, then so are '$(\text{—}\,1\,\text{—})$' and '$(\text{—}\,m'\,\text{—})$'.

If '$(\text{—}\,1\,\text{—})$' or '$(\text{—}\,m'\,\text{—})$' are theorems, then so is '$(\exists \mathfrak{n})(\text{—}\,\mathfrak{n}\,\text{—})$'.

These rules may evidently be derived from this axiom together with the theses

$$n = 1 \supset [(\text{—}\,n\,\text{—}) \equiv (\text{—}\,1\,\text{—})]$$
$$n = m' \supset [(\text{—}\,n\,\text{—}) \equiv (\text{—}\,m'\,\text{—})]$$

which will be immediate consequences of D3 for any of the contexts '$(\text{—}\,n\,\text{—})$' that actually appear in these proofs.

As regards other rules, I assume rules of first-order quantification theory without explicit citation, and cite rules proper to second-order quantification theory as 'Q'. Transitions depending on the ancestral are furnished with the citation 'A' followed by a numeral which refers to the relevant theorem of appendix 1 to chapter 5. (The citation 'Induction' also refers forward to the discussion of the ancestral in section 3 of that chapter.) All transitions depending on the fact that identity between numbers, as defined in D3, licenses substitution in whatever context is being considered are furnished simply with the citation 'D3'. Transitions depending on the fact that identity between objects is reflexive, symmetrical, and transitive are furnished with the citation 'ID'.

I give first proofs of the theorems cited in section 2 as obtainable 'without any dubious manœuvring':

THEOREM 1. $n \geqslant 1$

1.	$n=1 \vee n{*}\mathbf{S}1$	Ax 2, A7
2.	$n=1 \vee n\mathbf{S}1 \vee (\exists\mathfrak{m})(n\mathbf{S}\mathfrak{m} \mathbin{\&} \mathfrak{m}{*}\mathbf{S}1)$	1, A3
3.	$n=1 \vee (\exists\mathfrak{m})(n\mathbf{S}\mathfrak{m})$	2
4.	—$n\mathbf{S}m$	Hyp
5.	—$(\forall\mathfrak{f})((\exists nx)(\mathfrak{f}x) \equiv (\exists x)(\mathfrak{f}x \mathbin{\&} (\exists my)(\mathfrak{f}y \mathbin{\&} y \neq x)))$	4, D4, D2
6.	—$(\forall\mathfrak{f})((\exists nx)(\mathfrak{f}x) \supset (\exists x)(\mathfrak{f}x))$	5
7.	—$(\forall\mathfrak{f})((\exists nx)(\mathfrak{f}x) \supset (\exists 1x)(\mathfrak{f}x))$	6, D1
8.	—$n \geqslant 1$	7, D5
9.	$n\mathbf{S}m \supset n \geqslant 1$	4–8
10.	$n=1 \supset n \geqslant 1$	D3, D5
11.	$n \geqslant 1$	3, 9, 10

THEOREM 2. $n \geqslant m \supset n' \geqslant m'$

1.	—$n \geqslant m$	Hyp
2.	—$(\forall\mathfrak{f})((\exists nx)(\mathfrak{f}x) \supset (\exists mx)(\mathfrak{f}x))$	1, D5
3.	—$(\exists nx)(Fx \mathbin{\&} x \neq a) \supset (\exists mx)(Fx \mathbin{\&} x \neq a)$	2, Q
4.	—$(Fa \mathbin{\&} (\exists nx)(Fx \mathbin{\&} x \neq a)) \supset (Fa \mathbin{\&} (\exists mx)(Fx \mathbin{\&} x \neq a))$	3
5.	—$(\forall y)((Fy \mathbin{\&} (\exists nx)(Fx \mathbin{\&} x \neq y)) \supset (Fy \mathbin{\&} (\exists mx)(Fx \mathbin{\&} x \neq y)))$	1–4
6.	—$(\exists y)(Fy \mathbin{\&} (\exists nx)(Fx \mathbin{\&} x \neq y)) \supset (\exists y)(Fy \mathbin{\&} (\exists mx)(Fx \mathbin{\&} x \neq y))$	5
7.	—$(\exists n'x)(Fx) \supset (\exists m'x)(Fx)$	6, D2
8.	—$(\forall\mathfrak{f})((\exists n'x)(\mathfrak{f}x) \supset (\exists m'x)(\mathfrak{f}x))$	1–7
9.	—$n' \geqslant m'$	8, D5
10.	$n \geqslant m \supset n' \geqslant m'$	1–9

THEOREM 3. $1 \neq n'$

1.	—$1=n'$	Hyp
2.	—$1=n' \mathbin{\&} n \geqslant 1$	1, Th 1
3.	—$1=n' \mathbin{\&} n' \geqslant 1'$	2, Th 2
4.	—$1 \geqslant 1'$	3, D3
5.	—$(\forall\mathfrak{f})((\exists 1x)(\mathfrak{f}x) \supset (\exists 1'x)(\mathfrak{f}x))$	4, D5
6.	—$(\forall\mathfrak{f})((\exists x)(\mathfrak{f}x) \supset (\exists x)(\exists y)(\mathfrak{f}x \mathbin{\&} \mathfrak{f}y \mathbin{\&} x \neq y)$	5, D1–2
7.	—$(\exists x)(x=a) \supset (\exists x)(\exists y)(x=a \mathbin{\&} y=a \mathbin{\&} x \neq y)$	6, Q
8.	—$(\exists x)(\exists y)(x=y \mathbin{\&} x \neq y)$	7, ID
9.	$1 \neq n'$	1–8

THEOREM 4. $n \star Sm \supset (n = m' \vee n \star Sm')$

1.	— $n \star Sm$	Hyp
2.	— $nSm \vee (\exists \mathfrak{p})(n \star S\mathfrak{p} \,\&\, \mathfrak{p}Sm)$	1, A3, A6
3.	— $n = m' \vee (\exists \mathfrak{p})(n \star S\mathfrak{p} \,\&\, \mathfrak{p} = m')$	2, D3–4
4.	— $n = m' \vee n \star Sm'$	3, D3
5.	$n \star Sm \supset (n = m' \vee n \star Sm')$	1–4

THEOREM 5. $n \star Sm \vee n = m \vee m \star Sn$

1.	$n \star S1 \vee n = 1 \vee 1 \star Sn$	Ax2, A7
2.	— $n \star Sm \vee n = m \vee m \star Sn$	Hyp
3.	— $n \star Sm \vee m'Sn \vee (m'Sm \,\&\, m \star Sn)$	2, D4, D3
4.	— $n \star Sm \vee m' \star Sn$	3, A1
5.	— $n \star Sm' \vee n = m' \vee m' \star Sn$	4, Th 4
6.	$(n \star Sm \vee n = m \vee m \star Sn) \supset (n \star Sm' \vee n = m' \vee m' \star Sn)$	2–5
7.	$n \star Sm \vee n = m \vee m \star Sn$	1, 6, Induction

THEOREM 6. $n*Sm \supset n \geqslant m$

1.	$n*S1 \supset n \geqslant 1$	Th 1
2.	— $(\forall \mathfrak{n})(\mathfrak{n}*Sm \supset \mathfrak{n} \geqslant m)$	Hyp
3.	— — $p*Sm'$	Hyp
4.	— — $p*Sm' \,\&\, m'Sm$	3, D4
5.	— — $(\exists \mathfrak{n})(p*S\mathfrak{n} \,\&\, \mathfrak{n}\,Sm)$	4
6.	— — $(\exists \mathfrak{n})(pS\mathfrak{n} \,\&\, \mathfrak{n}*Sm)$	5, A9
7.	— — $(\exists \mathfrak{n})(pS\mathfrak{n} \,\&\, \mathfrak{n} \geqslant m)$	6, 2
8.	— — $(\exists \mathfrak{n})(pS\mathfrak{n} \,\&\, \mathfrak{n}' \geqslant m')$	7, Th 2
9.	— — $(\exists \mathfrak{n})(p = \mathfrak{n}' \,\&\, \mathfrak{n}' \geqslant m')$	8, D3–4
10.	— — $p \geqslant m'$	9, D3
11.	— $p*Sm' \supset p \geqslant m'$	3–10
12.	— $(\forall \mathfrak{n})(\mathfrak{n}*Sm' \supset \mathfrak{n} \geqslant m')$	2–11
13.	$(\forall \mathfrak{n})(\mathfrak{n}*Sm \supset \mathfrak{n} \geqslant m) \supset (\forall \mathfrak{n})(\mathfrak{n}*Sm' \supset \mathfrak{n} \geqslant m')$	2–12
14.	$n*Sm \supset n \geqslant m$	1, 13, Induction

As a preparation for the deduction of Frege's theorem I now prove two lemmas which are useful in counting. In the form in which I give them here, these also can be obtained 'without any dubious manœvring'.

LEMMA 1. $(\forall x)(Fx \supset Gx) \supset ((\exists nx)(Fx) \supset (\exists nx)(Gx))$

1.	$(\forall x)(Fx \supset Gx) \supset ((\exists 1x)(Fx) \supset (\exists 1x)(Gx))$	D1
2.	— $(\forall \mathfrak{f})(\forall \mathfrak{g})((\forall x)(\mathfrak{f}x \supset \mathfrak{g}x) \supset ((\exists nx)(\mathfrak{f}x) \supset (\exists nx)(\mathfrak{g}x)))$	Hyp
3.	— — $(\forall x)(Fx \supset Gx)$	Hyp
4.	— — $(\forall x)((Fx \,\&\, x \neq a) \supset (Gx \,\&\, x \neq a))$	3
5.	— — $(\exists nx)(Fx \,\&\, x \neq a) \supset (\exists nx)(Gx \,\&\, x \neq a)$	2, 4, Q
6.	— — $(Fa \,\&\, (\exists nx)(Fx \,\&\, x \neq a)) \supset (Ga \,\&\, (\exists nx)(Gx \,\&\, x \neq a))$	3, 5
7.	— — $(\exists n'x)(Fx) \supset (\exists n'x)(Gx)$	6, D2

8. $-(\forall x)(Fx \supset Gx) \supset ((\exists n'x)(Fx) \supset (\exists n'x)(Gx))$ — 3–7
9. $-(\forall \mathfrak{f})(\forall \mathfrak{g})((\forall x)(\mathfrak{f}x \supset \mathfrak{g}x) \supset ((\exists n'x)(\mathfrak{f}x) \supset (\exists n'x)(\mathfrak{g}x)))$ — 2–8
10. $(1\ \&\ (2 \supset 9)) \supset$ Prop. — Induction

Lemma 2.[16] $(\exists n'x)(Fx) \supset (\exists nx)(Fx\ \&\ x \neq a)$

1. $(\exists 1'x)(Fx) \supset (\exists 1x)(Fx\ \&\ x \neq a)$ — D1–2, ID
2. $-(\forall \mathfrak{f})((\exists n'x)(\mathfrak{f}x) \supset (\exists nx)(\mathfrak{f}x\ \&\ x \neq a))$ — Hyp
3. $-\,-Fb\ \&\ (\exists n'x)(Fx\ \&\ x \neq b)$ — Hyp
4. $-\,-Fb\ \&\ (\exists nx)(Fx\ \&\ x \neq b\ \&\ x \neq a)$ — 2, 3, Q
5. $-\,-b \neq a \supset (Fb\ \&\ b \neq a\ \&\ (\exists nx)(Fx\ \&\ x \neq a\ \&\ x \neq b))$ — 4
6. $-\,-b \neq a \supset (\exists n'x)(Fx\ \&\ x \neq a)$ — 5, D2
7. $-\,-\,-b = a$ — Hyp
8. $-\,-\,-(\forall x)((Fx\ \&\ x \neq b\ \&\ x \neq c) \supset (Fx\ \&\ x \neq a\ \&\ x \neq c))$ — 7, ID
9. $-\,-\,-(\exists nx)(Fx\ \&\ x \neq b\ \&\ x \neq c)$
   $\supset (\exists nx)(Fx\ \&\ x \neq a\ \&\ x \neq c)$ — 8, Lemma 1
10. $-\,-\,-(Fc\ \&\ c \neq b) \supset (Fc\ \&\ c \neq a)$ — 7, ID
11. $-\,-\,-(Fc\ \&\ c \neq b\ \&\ (\exists nx)(Fx\ \&\ x \neq b\ \&\ x \neq c))$
    $\supset (Fc\ \&\ c \neq a\ \&\ (\exists nx)(Fx\ \&\ x \neq a\ \&\ x \neq c))$ — 9, 10
12. $-\,-\,-(\exists n'x)(Fx\ \&\ x \neq b) \supset (\exists n'x)(Fx\ \&\ x \neq a)$ — 11, D2
13. $-\,-\,-(\exists n'x)(Fx\ \&\ x \neq a)$ — 3, 12
14. $-\,-b = a \supset (\exists n'x)(Fx\ \&\ x \neq a)$ — 7–13
15. $-\,-(\exists n'x)(Fx\ \&\ x \neq a)$ — 6, 14
16. $-(Fb\ \&\ (\exists n'x)(Fx\ \&\ x \neq b)) \supset (\exists n'x)(Fx\ \&\ x \neq a)$ — 3–15
17. $-(\exists n''x)(Fx) \supset (\exists n'x)(Fx\ \&\ x \neq a)$ — 16, D2
18. $-(\forall \mathfrak{f})((\exists n''x)(\mathfrak{f}x) \supset (\exists n'x)(\mathfrak{f}x\ \&\ x \neq a))$ — 2–17
19. $(1\ \&\ (2 \supset 18)) \supset$ Prop. — Induction

For the proof of Frege's theorem we now suppose that the previous definitions D1–D5 are rewritten with 'x' supplanted everywhere by the numerical variable '$\mathfrak{p}$', and 'F' and 'f' supplanted everywhere by 'Φ' and 'ϕ' as schematic letter and variable for properties of numbers. Axioms 1–2 are retained with the new definitions. It will be seen that this does not affect the proofs of the previous theorems, for these employed only quantification theory and the fact that identity is a reflexive, symmetrical, and transitive relation, which is equally true of the identity now defined by D3. With these alterations the proof of Frege's theorem and of the missing postulate can be given as follows.

Theorem 7. $(\exists n\mathfrak{m})(n * S\mathfrak{m})$

1. $1 * S1$ — D3, A7
2. $(\exists 1\mathfrak{m})(1 * S\mathfrak{m})$ — 1, D1
3. $n'Sn$ — D4
4. $(\forall \mathfrak{m})(n * S\mathfrak{m} \supset n' * S\mathfrak{m})$ — 3, A8
5. $(\exists n\mathfrak{m})(n * S\mathfrak{m}) \supset (\exists n\mathfrak{m})(n' * S\mathfrak{m})$ — 4, Lemma 1

[16] In view of the substitutivity of identity the subdeduction in lines 7–12 may appear superfluous here. But I wished to show that this proof depends only on the symmetry and transitivity of identity, for this is of some importance later.

6.	$-n \geqslant n'$	Hyp
7.	$-(\exists n\mathfrak{m})(n'*\mathbf{S}\mathfrak{m}) \supset (\exists n'\mathfrak{m})(n'*\mathbf{S}\mathfrak{m})$	6, D5
8.	$-(\exists n\mathfrak{m})(n*\mathbf{S}\mathfrak{m}) \supset (\exists n'\mathfrak{m})(n'*\mathbf{S}\mathfrak{m})$	5, 7
9.	$n \geqslant n' \supset ((\exists n\mathfrak{m})(n*\mathbf{S}\mathfrak{m}) \supset (\exists n'\mathfrak{m})(n'*\mathbf{S}\mathfrak{m}))$	6–8
10.	$-\sim n \geqslant n'$	Hyp
11.	$-\sim n*\mathbf{S}n'$	10, Th 6
12.	$-(\forall\mathfrak{m})(n*\mathbf{S}\mathfrak{m} \supset \mathfrak{m} \neq n')$	11, D3
13.	$-(\forall\mathfrak{m})(n*\mathbf{S}\mathfrak{m} \supset (n'*\mathbf{S}\mathfrak{m} \,\&\, \mathfrak{m} \neq n'))$	4, 12
14.	$-(\exists n\mathfrak{m})(n*\mathbf{S}\mathfrak{m}) \supset (\exists n\mathfrak{m})(n'*\mathbf{S}\mathfrak{m} \,\&\, \mathfrak{m} \neq n')$	13, Lemma 1
15.	$-(\exists n\mathfrak{m})(n*\mathbf{S}\mathfrak{m}) \supset (n'*\mathbf{S}n' \,\&\, (\exists n\mathfrak{m})(n'*\mathbf{S}\mathfrak{m} \,\&\, \mathfrak{m} \neq n'))$	14, A 7
16.	$-(\exists n\mathfrak{m})(n*\mathbf{S}\mathfrak{m}) \supset (\exists n'\mathfrak{m})(n'*\mathbf{S}\mathfrak{m})$	15, D2
17.	$\sim n \geqslant n' \supset ((\exists n\mathfrak{m})(n*\mathbf{S}\mathfrak{m}) \supset (\exists n'\mathfrak{m})(n'*\mathbf{S}\mathfrak{m}))$	10–16
18.	$(\exists n\mathfrak{m})(n*\mathbf{S}\mathfrak{m}) \supset (\exists n'\mathfrak{m})(n'*\mathbf{S}\mathfrak{m})$	9, 17
19.	$(\exists n\mathfrak{m})(n*\mathbf{S}\mathfrak{m})$	2, 18, Induction

THEOREM 8. $\sim(\exists n'\mathfrak{m})(n*\mathbf{S}\mathfrak{m})$

1.	$-(\exists 1'\mathfrak{m})(1*\mathbf{S}\mathfrak{m})$	Hyp
2.	$-(\exists\mathfrak{n})(\exists\mathfrak{m})(1*\mathbf{S}\mathfrak{n} \,\&\, 1*\mathbf{S}\mathfrak{m} \,\&\, \mathfrak{n} \neq \mathfrak{m})$	1, D1–2
3.	$-(\exists\mathfrak{n})(\exists\mathfrak{m})(1*\mathbf{S}\mathfrak{n} \,\&\, \mathfrak{n}*\mathbf{S}1 \,\&\, 1*\mathbf{S}\mathfrak{m} \,\&\, \mathfrak{m}*\mathbf{S}1 \,\&\, \mathfrak{n} \neq \mathfrak{m})$	2, Ax2
4.	$-(\exists\mathfrak{n})(\exists\mathfrak{m})(1 \geqslant \mathfrak{n} \,\&\, \mathfrak{n} \geqslant 1 \,\&\, 1 \geqslant \mathfrak{m} \,\&\, \mathfrak{m} \geqslant 1 \,\&\, \mathfrak{n} \neq \mathfrak{m})$	3, Th 6
5.	$-(\exists\mathfrak{n})(\exists\mathfrak{m})(1 = \mathfrak{n} \,\&\, 1 = \mathfrak{m} \,\&\, \mathfrak{n} \neq \mathfrak{m})$	4, D5, D3
6.	$-(\exists\mathfrak{n})(\exists\mathfrak{m})(\mathfrak{n} = \mathfrak{m} \,\&\, \mathfrak{n} \neq \mathfrak{m})$	5, D3
7.	$\sim(\exists 1'\mathfrak{m})(1*\mathbf{S}\mathfrak{m})$	1–6
8.	$-n'*\mathbf{S}m \,\&\, m \neq n'$	Hyp
9.	$-n' \geqslant m \,\&\, m \neq n'$	8, Th 6
10.	$-\sim m \geqslant n'$	9, D5, D3
11.	$-\sim m*\mathbf{S}n'$	10, Th 6
12.	$-\sim m\star\mathbf{S}n$	11, Th 4, A7
13.	$-n*\mathbf{S}m$	12, Th 5, A7
14.	$(n'*\mathbf{S}m \,\&\, m \neq n') \supset n*\mathbf{S}m$	8–13
15.	$(\forall\mathfrak{m})((n'*\mathbf{S}\mathfrak{m} \,\&\, \mathfrak{m} \neq n') \supset n*\mathbf{S}\mathfrak{m})$	14
16.	$(\exists n'\mathfrak{m})(n'*\mathbf{S}\mathfrak{m} \,\&\, \mathfrak{m} \neq n') \supset (\exists n'\mathfrak{m})(n*\mathbf{S}\mathfrak{m})$	15, Lemma 1
17.	$(\exists n''\mathfrak{m})(n'*\mathbf{S}\mathfrak{m}) \supset (\exists n'\mathfrak{m})(n*\mathbf{S}\mathfrak{m})$	16, Lemma 2
18.	$\sim(\exists n'\mathfrak{m})(n*\mathbf{S}\mathfrak{m}) \supset \sim(\exists n''\mathfrak{m})(n'*\mathbf{S}\mathfrak{m})$	17
19.	$\sim(\exists n'\mathfrak{m})(n*\mathbf{S}\mathfrak{m})$	7, 18, Induction

THEOREM 9. $n' = m' \supset n = m$

1.	$-n*\mathbf{S}n'$	Hyp
2.	$-n \geqslant n'$	1, Th 6
3.	$-(\exists n\mathfrak{m})(n*\mathbf{S}\mathfrak{m}) \supset (\exists n'\mathfrak{m})(n*\mathbf{S}\mathfrak{m})$	2, D5
4.	$\sim n*\mathbf{S}n'$	1–3, Th 7–8
5.	$n' = m' \supset \sim n*\mathbf{S}m'$	4, D3
6.	$n' = m' \supset \sim n\star\mathbf{S}m$	5, Th 4, A7
7.	$m' = n' \supset \sim m\star\mathbf{S}n$	6
8.	$n' = m' \supset n = m$	6, 7, Th 5

2. Objects

1. Russell's Theory of Types

THE question whether numbers are objects is obviously not to be taken as the question whether numbers are *physical* objects (roughly, objects with spatio-temporal location) nor yet as the question whether numbers are *mental* objects (as mental images perhaps are). For I take it that to both these questions the answer is quite evidently 'no'. Rather, the question we are concerned with is the question whether numbers are *logical* objects, or—more clearly put—whether numbers are of the *logical type of objects*, and to elucidate this question we must accordingly begin upon an elucidation of the theory of logical types.

The need for a theory of types may well be brought out by considering certain paradoxes, notably Russell's paradox, and this is the way in which Russell himself generally sets about it. However, I do not at all agree with the main conclusion that Russell draws from his paradoxes, namely that certain uses of the word 'all' should be regarded as meaningless, so I shall begin by considering Russell's position on this.

In his article *Mathematical Logic as based on the Theory of Types* Russell first reviews a series of paradoxes and then concludes:

> All our contradictions have in common the assumption of a totality such that, if it were legitimate, it would at once be enlarged by new members defined in terms of itself. This leads us to the rule: 'Whatever involves *all* of a collection must not be one of the collection'; or, conversely: 'If, provided a certain collection had a total, it would have members only definable in terms of that total, then the said collection has no total'. And when I say that a collection has no total, I mean that statements about *all* its members are nonsense.
>
> (*Logic & Knowledge*, p. 63)

The structure of Russell's argument here is: it *seems* to be possible to define new members of certain collections by speaking of all members of those collections, but in certain cases the new members so defined would have contradictory properties. Since, then, there cannot be such new members, it never is possible to define new members by speaking of all members, and therefore it never makes sense to speak of all members. A moment's reflection, however, will make it clear that the conclusion to this argument goes far beyond what is required by its premiss. In the first place, wherever Russell finds that paradoxes arise from *some* attempts to define new members by speaking of all members, he concludes that *all* attempts to define new members by speaking of all members are illegitimate. And in the second place he moves without further argument from the position that we cannot *define new members* by speaking of all members to the position that we cannot do *anything* by speaking of all members, i.e. that we cannot speak of all members at all. It would certainly seem that this second step is quite gratuitous; as Russell has presented the argument, it appears that all that we shall require to prevent the paradoxes arising is a restriction on admissible *definitions*, and not a general doctrine of the meaninglessness of certain universal quantifications.

I think it is a significant fact that a large number of the paradoxes Russell discusses are very naturally presented as concerned in one way or another with definitions.[1] For instance in this same article Russell puts the paradox that is called after him in the form

> Let w be the class of all those classes which are not members of themselves. Then, whatever class x may be, 'x is a w' is equivalent to 'x is not an x'. Hence, giving to x the value w, 'w is a w' is equivalent to 'w is not a w'. (*Logic & Knowledge*, p. 59)

This way of putting the paradox is entirely natural, and very common, and we may notice that the opening sentence is almost certain to be taken as an attempted *definition* of w. It may seem that there is no appeal to definition here, since what is being pointed out is just that there is no true proposition of the sort

$$(\forall x)(x \in w \equiv \sim x \in x)$$

[1] A conspicuous exception is the paradox of the Liar; I shall not attempt to deal with that paradox here.

where the quantifier '$(\forall x)(—x—)$' has w in its range, since any such proposition will imply the contradiction

$$w \in w \equiv \sim w \in w.$$

And in a sense this *is* all that is being pointed out. But why should such a point be thought to be in any way paradoxical? Surely at least part of the answer here is that we *also* seem to have a reason for supposing that there *must* be a true proposition of the sort in question, namely that we can produce one which is *true by definition*. For classes are defined by their membership, and since this proposition states the condition for being a member of w it can be viewed simply as the definition of w, and therefore it *must* be true. Generally, then, what makes a paradox of this sort paradoxical is at least partly that it consists of a proof of the falsehood of some proposition which we would otherwise have believed to be true by definition, and an adequate 'solution' of the paradox must therefore explain what was wrong with our reasons for believing it to be true by definition. Consequently what needs investigation is primarily the concept of *definition*, and there is no reason to suppose that our findings will have implications for the correct employment of quantification in propositions that are not definitions.

This point will bear some elaboration. In Russell's opinion what is wrong with the attempted definition just discussed is that it is *impredicative*, i.e. that it is an attempt to define a thing of a certain kind (a class) by means of a quantification over all things of that kind (all classes); and he goes on to explain that what is wrong with an impredicative definition is that it is covertly *circular*, for if a thing is defined in terms of all things of the same kind as it then it is—*inter alia*—defined in terms of itself. But circularity is precisely a defect of *definitions*, and is totally irrelevant to anything that is not offered as a definition. Of course it may sometimes happen that propositions embodying unacceptable definitions are themselves unacceptable propositions, and this was the case with the example we considered, but there is no reason to suppose that it is always the case. Consider, for instance, the commonly accepted definition of identity

$$\text{`}a = b\text{'} \quad \text{for} \quad \text{`}(\forall \mathfrak{f})(\mathfrak{f}a \equiv \mathfrak{f}b)\text{'}.$$

This definition has the result that any monadic predicate of the form '... $= b$' is defined by means of a quantification over all monadic predicates, and Russell accordingly proscribes it as illegitimate.[2] Now it may perhaps be that the impredicative nature of this definition should lead us to reject it *as* a definition, just as the evident circularity of the attempted definition:

$$\text{`a=b'} \quad \text{for} \quad \text{`}(\forall x)(x=a \equiv x=b)\text{'}$$

would presumably lead us to reject this as a definition. But quite apart from the question whether we have here an acceptable *definition* we can simply ask whether we even have a *truth*—i.e. whether the corresponding propositions

$$\vdash a=b \equiv (\forall \mathfrak{f})(\mathfrak{f}a \equiv \mathfrak{f}b)$$

$$\vdash a=b \equiv (\forall x)(x=a \equiv x=b)$$

are even true. I take it that nobody will raise any doubts about the truth of the second, although it corresponds to a clearly circular definition, but Russell does object to the first precisely on this ground. Since the corresponding definition would be impredicative, the first proposition here is one that Russell rejects as nonsense!

As a matter of fact I do not think that definitions which are impredicative in Russell's sense must always be rejected on the score of circularity. *One* way of regarding a definition is as a merely conventional abbreviation—for instance '$a \neq b$' is a merely conventional abbreviation of '$\sim(a=b)$'—and in this way of looking at the matter we may properly say that it is *signs* that are defined. Now a *sign* may be said to be adequately defined if and only if it *can* be seen merely as yielding an abbreviation, and this will be so if and only if the definition gives a procedure for *eliminating* that sign from any context in which it occurs. By this criterion the attempted definition of 'w' (the Russell class) fails, because evidently the definition does not give us any procedure for eliminating 'w' from the context '$w \in w$'; but on the other hand the definition of '=' is completely successful. Regarding this latter definition merely as a conventional abbreviation there is no conceivable ground for objecting to it; it does not enable us to say, or to prove, anything that could not be said or proved without it. (In this connection it may be noted (i) that a sign does not have to be defined for *all* contexts but only

[2] See his introduction to *Principia Mathematica*, chapter 2, sections v–vi.

for those in which we wish to use it; (ii) that we may well give different definitions for the sign as it occurs in different contexts (as Aristotle noticed long ago apropos of 'healthy man' and 'healthy climate'); and (iii) that the elimination may well require *several* applications of the rule given. For instance the pair of recursive equations

$$n+1=n',$$

$$n+m'=(n+m)'$$

will allow us to eliminate the sign '+' from any context in which it occurs between signs of the form '$1''\cdots'$', provided we apply the second equation a sufficient number of times, and they may therefore be regarded as yielding an adequate definition of the sign '+' *as it occurs in these contexts*.)

Now in the more interesting cases it is clearly oversimplifying matters to say that a definition is a mere abbreviation. This may be seen, for instance, by reflecting that one might wish to claim that the definition of the sign '=' just discussed *is* the definition of *identity*, and if so one would be claiming that we have here the correct definition of an already familiar *concept*, and not just the arbitrary introduction of a new *sign*. What exactly this way of regarding definition comes to is a matter of some difficulty, and I shall not discuss it until chapter 6. Again I do not think that being impredicative is necessarily a defect of definitions so regarded, but that is rather a side-issue from the main point that I want to make here, which is that whatever the truth about impredicative definitions Russell has given us no sound reason for suspecting the meaningfulness of any universal quantification.

Indeed Russell's own attempts to dispense with such quantification lead him into quite intolerable difficulties, as may be seen by continuing our inquiry into the proposition

$$\vdash a=b \equiv (\forall \mathfrak{f})(\mathfrak{f}a \equiv \mathfrak{f}b).$$

This proposition Russell ostensibly rejects as *senseless*, and yet in fact it seems clear that he is committed to its *truth*. For, believing that quantification over *all* properties was not legitimate, Russell defined identity by saying that objects were to be identical if and only if they had all their *first-order* properties in common, explaining that being identical with something was not itself a first-order property but a second-order property. (In general, properties defined by means of a quantification over properties of n'th order are themselves of (n+1)'th order.) However, this leaves it as a formal possibility that objects which are identical according to this definition might yet differ in their higher-order properties, which Russell clearly found unacceptable, and for this and other reasons he found himself bound to introduce the axiom of reducibility according to which any property of any order is equivalent to some first-order property. But this axiom at once yields the consequence that objects have all their first-order properties in common if and only if they have *all* their properties in common, and so we find that Russell's own definition of identity, together with his axiom of reducibility, enables us to deduce the very proposition he had initially condemned as nonsensical.

I say that we can deduce the allegedly nonsensical proposition, though I should admit that the deduction cannot actually be carried through in Russell's system, since that system does not contain the theorem that all first-order properties are indeed properties. For any way of saying this must issue in some statement such as 'whatever holds for *all* properties holds also for all first-order properties', and so it must involve a quantification over all properties, which Russell provides no means of expressing since he regards it as illegitimate. But in that case, one will ask, how does Russell manage to express his axiom of reducibility, which again involves a quantification over all properties in so far as it states that *all* properties are equivalent to first-order properties? In answer to this Russell propounds a distinction between 'all' and 'any' by which 'any property' is a legitimate expression though 'all properties' is not, and he then puts his axiom in the form '*any* property is equivalent to some first-order property'. But it seems to me that this distinction is entirely spurious. For if 'any property is equivalent to some first-order property' is a proposition with a perfectly good sense, then surely its negation will also be a proposition with a perfectly good sense, but its negation will be expressed as 'not *all* properties are equivalent to first-order properties' and this cannot be written with 'any' in place of 'all'. In fact Russell's distinction between 'all' and 'any' just is the distinction between a universal quantifier which occurs at the beginning of a proposition and one that occurs anywhere else in a proposition,[3] and I do not see how this could make the difference between sense and nonsense. If '$(\forall \mathfrak{f})(-\mathfrak{f}-)$' makes sense, then so does '$\sim(\forall \mathfrak{f})(-\mathfrak{f}-)$'; and consequently either Russell's axiom of reducibility, which he puts forward as true, is nonsense, or the proposition concerning identity, which he rejects as nonsense, is on his assumptions true.

This is just one instance of the difficulties which Russell was led into by his decision to regard certain quantifications as meaningless, and further instances could certainly be given.[4] Further, we have seen that the paradoxes from which he starts require no such general doctrines of meaninglessness, but simply the need for a cautious attitude to what can be done by definition. So it begins to appear that the whole Russellian theory of types is entirely superfluous. But that, I think, would be going too far.

2. The Principle of Abstraction

It was pointed out by Ramsey[5] that Russell's theory of types could be split into two parts: first the *simple* theory of types which distinguishes individuals, properties of individuals, properties of properties of individuals, and so on, into distinct types; and second the *ramified* theory which distinguishes within each type (except the lowest) the properties of first order, second order, third order, and so on. In Ramsey's view there was no need to adopt the complications of the ramified theory, and it is essentially the ramified theory which I have so far been arguing against, since it is that part of the theory that proscribes quantification over properties of all orders at once. In this I have been quite fair to Russell, for Russell explicitly bases his whole theory on the principle quoted on p. 26 (the 'vicious-circle' principle), and this principle does yield the ramifications of the ramified theory in a fairly straightforward way, though it yields the simple theory only in conjunction with the very dubious premiss that a propositional function *involves* or *is only definable in terms*

[3] Russell speaks rather of the distinction between 'real' and 'apparent' variables; I regard this as simply another way of putting the distinction I mention, but I am here assuming the argument I develop more fully in the next chapter.

[4] See e.g. Pap, 'Types and Meaninglessness', *Mind*, 1960.

[5] 'Foundations of Mathematics', especially sections 2–3 (in F. P. Ramsey, *Foundations of Mathematics and other Logical Essays*, ed. Braithwaite).

of all its values.[6] But, leaving this dubious premise on one side, we ought, I think, to consider the simple theory of types more closely.

Let us start once again from a consideration of Russell's own paradox. Using 'α' and 'β' as class-variables, we have seen that this paradox is just the point that the proposition

$$(\exists\beta)(\forall\alpha)(\alpha \in \beta \equiv {\sim}\alpha \in \alpha)$$

is false. It follows that we cannot accept what is generally known as the 'principle of abstraction', namely

$$\vdash (\exists\beta)(\forall\alpha)(\alpha \in \beta \equiv \text{—}\alpha\text{—})$$

(where '$\text{—}\alpha\text{—}$' is a schematic expression for any proposition concerning the class α), provided of course that we do recognize '${\sim}\alpha \in \alpha$' *as* a proposition concerning the class α. Russell, however, did not grant this proviso, and preferred to say that it made no sense to suppose that a class was or was not a member of itself, though with an *ordinary* understanding of class-expressions this would seem to make perfectly good sense. Intuitively speaking, it seems quite correct to say that the class of men is not itself a man, and that the class of classes which have some members does itself have some members, and since '. . . is a so-and-so' is ordinarily thought to be interchangeable with '. . . is a member of the class of so-and-so's' it would therefore follow that the class of men is not a member of itself and that the class of classes which have members is a member of itself. I say that these propositions are ones that we should intuitively regard as true, though I do not positively affirm that they are true since there may well be ground for saying that *there is no such class as* the class of all classes which have members, and I suppose there might be ground for saying that *there is no such class as* the class of all men. But even if these propositions fail to be true because the classes in question fail to exist, I do not see that there is any plausibility in the view that they are without sense; it will simply be that the definite descriptions 'the class of so-and-so's' lack a reference, and that malady would leave the sense of the propositions quite intact.

I do not urge this point in criticism of Russell's own system. for in that system the so-called class-expressions '$\hat{x}(\text{—}x\text{—})$' and

[6] Introduction to *Principia Mathematica*, chapter II, section (ii).

class-variables 'α', 'β', are *not* to be understood in the *ordinary* way at all. Russell often characterizes his theory as a 'no-class' theory on the ground that it does not anywhere assert the existence of classes, and I believe that this characterization is perfectly correct. For in his system class-expressions are given a contextual definition which is roughly of the form[7]

$$\text{`}(\ldots \hat{x}(-x-)\ldots)\text{'} \quad \text{for} \quad \text{`}(\exists \mathfrak{f})((\forall x)(\mathfrak{f}x \equiv -x-) \ \& \ \ldots \mathfrak{f} \ldots)\text{'}$$

and this makes it quite clear that Russell's class-expressions occupy contexts available to *predicate*-variables, and so have the grammatical form not of subject-expressions but of predicate-expressions. And with this understanding of class-expressions a concatenation of symbols such as '$\alpha \in \alpha$' is indeed quite meaningless. As Russell shows,[8] with his definition of 'classes' there could be propositions of the form '$\alpha \in \alpha$' only if there could be propositions of the form 'F(F)', which evidently there cannot be. A predicate-letter has a use only in conjunction with a subject-letter or a subject-variable, and there is an important sense in which just this is the heart of the whole theory of types.

Now it may be objected that this is entirely arbitrary stipulation. For surely it is easy to bring examples of propositions which we might wish to recognize as of the form 'F(F)', including perhaps some of:

> To be loved by all men is loved by all men
> Abstractness is an abstract property
> The property of being seldom heeded is seldom heeded
> The concept of a vague concept is a vague concept
> '. . . expresses a predicate' expresses a predicate

and obviously other examples could be suggested. Now of course I do not want to say that it simply is not possible to give the symbol 'F(F)' a use in which it would represent the form of some of these examples, or others like them, but I do want to say that we cannot recognize the two occurrences of 'F' in this symbol as doing the same job, and therefore the symbol must be misleading. For in its ordinary use the letter 'F' may be thought of as

[7] Naturally Russell's own version is a little more complicated, for the quantifier '$(\exists \mathfrak{f})(-\mathfrak{f}-)$' is not available to him. But the differences are of no importance here.

[8] Introduction to *Principia Mathematica*, chapter III, p. 79. (And compare the supporting argument in chapter II section (iv), which I can endorse while rejecting the argument in chapter II section (ii).)

standing for what is left of a proposition when its subject is dropped out, and—to employ a familiar distinction—this will be a predicate *in use* and not a *mention* of a predicate. Evidently a proposition is expressed by *using* a predicate of a subject, and not by simply mentioning a predicate and mentioning a subject alongside: propositions cannot be viewed merely as *lists*, for (to bring just one objection) the difference between the forms 'aRb' and 'bRa' would then be quite inexplicable. So it must be recognized that if we are to use the symbol 'F(F)' the first 'F' stands in place of a predicate in use and the second 'F' stands in place of a mention of that predicate, which is why I say that the two 'F''s do different jobs so that the symbol is misleading.

It may seem, however, that the difference I have pointed out is not of very great consequence since the switch from use to mention is perfectly well understood. One would certainly suppose that whatever can be used can be mentioned, and indeed we have a variety of methods of forming grammatical subject-expressions for just this purpose—for instance by nominalizing adjectives with a suffix such as '-ness' or '-ity' or '-hood', or by nominalizing verb-phrases with gerundive or infinitive constructions, and perhaps adding such prefixes as 'the characteristic of . . .', 'the property of . . .', 'the concept of . . .', and so on; or again by taking an expression for a predicate in use and enclosing it in quotation marks (again with various prefixes); or in several other ways. Further, one would certainly suppose that whatever can be said by using a given predicate can equally well be said differently by mentioning that predicate and using instead a relation expressed by '. . . exemplifies . . .', '. . . is an instance of . . .', '. . . characterizes . . .', '. . . applies to . . .', '. . . is true of . . .', and so on. Let us now take over the familiar class-symbols '$\hat{x}$(—x—)' to represent whatever is the preferred way of mentioning the predicate used of anything x when it is said that —x—, with '... $\in$ ...' as the associated relation from the family just indicated. Then the principle which I said 'one would certainly suppose true' is that

$$\vdash(\forall x)(Fx \equiv x \in \hat{x}(Fx)).$$

But of course the trouble is that if the expressions here supplanted by '$\hat{x}$(Fx)' are grammatical subject-expressions, and are used to mention something, then they would seem to be logical subject-expressions, so this principle would imply

$$\vdash(\exists y)(\forall x)(Fx \equiv x \in y).$$

But *this* is the principle of abstraction which Russell observed to be contradictory.

What we must realize, then, is that the switch from use to mention is *not* after all so well understood for it appears to lead us straight into a contradiction. Now the contradiction depends on our being able to construe the expressions '$\hat{x}(-x-)$' as *subject*-expressions only in the sense that they are expressions used to mention or refer to *something or other*, for whatever it is that these expressions are taken as mentioning we may always interpret our variables 'x', 'y',..., as ranging over just those things, and so we cannot hope to evade the difficulty by restricting the notion of a logical subject so as to exclude, say, the traditional category of 'universals'. And if that part of the argument must be allowed to stand, the only further assumption that we need to obtain the contradiction seems to be the assumption that '$\sim\hat{x}(-x-) \in \hat{x}(-x-)$' makes sense, so that there is a predicate '$\sim ... \in ...$' which can in its turn be mentioned. One might perhaps suspect that it is not so much the mentioning of predicates that is causing the difficulty as the assumption that there is such a relation as '$... \in ...$' is supposed to be, for philosophers have very often urged that exemplification and its kin are very peculiar relations; but in this connection we should recall Frege's version of the paradox which seems to make no use of the alleged relation '$... \in ...$'. For Frege shows that we can still deduce a contradiction from the simple assumptions (i) that '$\hat{x}(Fx)$' always denotes something, (ii) that '$F(\hat{x}(Fx))$' always makes sense, and (iii) that there is a reflexive relation (which I symbolize by '$... \approx ...$') which holds between $\hat{x}(Fx)$ and $\hat{x}(Gx)$ only when $(\forall x)(Fx \equiv Gx)$.[9] (We may for instance take '$\hat{x}(Fx) \approx \hat{x}(Gx)$' as asserting that $\hat{x}(Fx)$ and $\hat{x}(Gx)$ are equivalent—or indeed identical—predicates.)

Now by our assumptions there clearly are propositions of the form '$(\forall \mathfrak{f})[\hat{x}(\mathfrak{f}x) \approx a \supset \sim\mathfrak{f}a]$', and these propositions make use of the predicate '$(\forall \mathfrak{f})[\hat{x}(\mathfrak{f}x) \approx ... \supset \sim\mathfrak{f}...]$', which is in turn mentioned by the expression '$\hat{y}(\forall \mathfrak{g})(\hat{x}(\mathfrak{g}x) \approx y \supset \sim\mathfrak{g}y)$'.[10] To suppose

[9] *The Basic Laws of Arithmetic*, vol. ii, Appendix. See in particular pp. 133–8 in Furth's translation.

[10] I use distinct variables 'y' and '$\mathfrak{g}$' only for clarity in the ensuing argument.

that this predicate applies to itself, i.e. that its substitution for 'F' in '$F(\hat{x}(Fx))$' yields a truth, is to suppose that

$$(\forall \mathfrak{f})(\hat{x}(\mathfrak{f}x) \approx \hat{y}(\forall \mathfrak{g})(\hat{x}(\mathfrak{g}x) \approx y \supset \sim \mathfrak{g}y) \supset \sim \mathfrak{f}(\hat{y}(\forall \mathfrak{g})(\hat{x}(\mathfrak{g}x) \approx y \supset \sim \mathfrak{g}y))).$$

But it turns out that this supposition is contradictory, as may be seen by noticing that it implies the special case in which the universally quantified variable '$\mathfrak{f}$' is taken as that very predicate that we began with. For the antecedent to the conditional then becomes

$$\hat{y}(\forall \mathfrak{g})(\hat{x}(\mathfrak{g}x) \approx y \supset \sim \mathfrak{g}y) \approx \hat{y}(\forall \mathfrak{g})(\hat{x}(\mathfrak{g}x) \approx y \supset \sim \mathfrak{g}y),$$

which must be true if '$\ldots \approx \ldots$' has the meaning we intend, while the consequent becomes

$$\sim(\forall \mathfrak{f})(\hat{x}(\mathfrak{f}x) \approx \hat{y}(\forall \mathfrak{g})(\hat{x}(\mathfrak{g}x) \approx y \supset \sim \mathfrak{g}y) \supset \sim \mathfrak{f}(\hat{y}(\forall \mathfrak{g})(\hat{x}(\mathfrak{g}x) \approx y \supset \sim \mathfrak{g}y)))$$

which is precisely the negation of our supposition. We have established, then, that our predicate does not apply to itself, and hence that

$$(\exists \mathfrak{f})(\hat{x}(\mathfrak{f}x) \approx \hat{y}(\forall \mathfrak{g})(\hat{x}(\mathfrak{g}x) \approx y \supset \sim \mathfrak{g}y) \;\&\; \mathfrak{f}(\hat{y}(\forall \mathfrak{g})(\hat{x}(\mathfrak{g}x) \approx y \supset \sim \mathfrak{g}y))).$$

But this result is already in conflict with our initial assumptions, as may be seen more clearly if we now introduce the abbreviation '(M...)' for our predicate '$(\forall \mathfrak{f})[\hat{x}(\mathfrak{f}x) \approx \ldots \supset \sim \mathfrak{f} \ldots]$'. For the result just stated clearly abbreviates to

$$(\exists \mathfrak{f})(\hat{x}(\mathfrak{f}x) \approx \hat{x}(Mx) \;\&\; \mathfrak{f}(\hat{x}(Mx))),$$

and this in turn abbreviates once more to

$$\sim M(\hat{x}(Mx))$$

(which is how we originally put the result). But putting these last two formulae together, and applying existential generalization, we deduce

$$(\exists \mathfrak{f})(\exists \mathfrak{g})(\hat{x}(\mathfrak{f}x) \approx \hat{x}(\mathfrak{g}x) \;\&\; \mathfrak{f}(\hat{x}(\mathfrak{g}x)) \;\&\; \sim \mathfrak{g}(\hat{x}(\mathfrak{g}x))),$$

whence

$$(\exists \mathfrak{f})(\exists \mathfrak{g})(\hat{x}(\mathfrak{f}x) \approx \hat{x}(\mathfrak{g}x) \;\&\; \sim(\forall x)(\mathfrak{f}x \equiv \mathfrak{g}x))$$

contradicting our original assumption (iii).

It may perhaps be pointed out that while the Fregean version of the paradox avoids the relation '$\ldots \in \ldots$' it does make use instead of a predicate expressed by quantifying over all predicates, and so would be prevented if we were to reinstate Russell's so-called 'vicious-circle' principle. However, I have already argued in section 1 that this principle is unacceptable, and so I shall not consider this escape further here. Reviewing these two versions of the paradox, then, we find that the assumptions common to both of them are

(i) that '$\hat{x}(Fx)$' always denotes something
(ii) that '$F(\hat{x}(Fx))$' always makes sense

(provided of course that 'F...' is a predicate with a perfectly good sense). The first version assumes in addition that there is a relation '... ∈ ...' such that

(iii*a*) $\vdash (\forall x)(Fx \equiv x \in \hat{x}(Fx))$,

but the second assumes only what is a weaker consequence of this,[11] namely that there is a relation '... ≈ ...' such that

(iii*b*) $\vdash \hat{x}(Fx) \approx \hat{x}(Fx)$
$\vdash (\hat{x}(Fx) \approx \hat{x}(Gx)) \supset (\forall x)(Fx \equiv Gx)$.

And I would think that there is no plausibility whatever in dropping this last assumption but retaining the original two—provided that we do continue to construe the expressions '$\hat{x}(—x—)$' as expressions that mention predicates.

For I should remark that the arguments we have been going through are entirely formal, and may be applied to quite different interpretations of the symbols '$\hat{x}(—x—)$'; in fact these symbols may be taken as representing *any* way in which an expression for a predicate in use may be transformed into a grammatical subject-expression used ostensibly to mention something. Obviously one relevant interpretation is the ordinary class-interpretation, by which prefixing '$\hat{x}$(. . .)' is taken as a representation of prefixing 'the class of all things which . . .'; another relevant interpretation is what one might call the *expression*-interpretation, by which prefixing '$\hat{x}$(. . .)' is taken as a representation of what is got by enclosing the predicate-expression in quotation marks and prefixing 'the (predicate-) expression . . .'; and for both of these interpretations our untutored intuitions would probably count all of the inconsistent assumptions true. A quite different interpretation is one in which '$\hat{x}$(. . .)' is taken as the familiar definite description operator 'the

[11] If we strengthen the implication in (iii*b*) to an equivalence, then (iii*b*) and (iii*a*) become interdeducible.

sole thing which . . .', and with this interpretation assumption (ii) is clearly true, while assumption (i) is true only if we adopt some arbitrary Fregean stipulation, in which case assumption (iii) is false. (On a Russellian view, of course, (i) is false anyway.) This case, then, gives rise to no puzzles, but the class-interpretation and the expression-interpretation seem to be very much on a par with our original.

I shall take it, then, that either assumption (i) or assumption (ii) must be rejected for these interpretations, and my preference is definitely for rejecting assumption (i). In the case of the class-interpretation I have already hinted that this course is to be preferred on p. 33, and it is I think the course that is almost always pursued nowadays. It simply is not in the least bit plausible to deny *sense* to such obvious truths as 'the class of men is not itself a man', but it is perfectly plausible to say that the form of words 'the class of so-and-so's' should be construed on the model of an ordinary definite description, and so may turn out to denote nothing even though it has a perfectly good sense. Just as ordinary definite descriptions may be introduced by a contextual definition of the sort

$$\text{`}(\ldots(\iota x)(\text{—}x\text{—})\ldots)\text{'} \quad \text{for} \quad \text{`}(\exists y)((\forall x)(\text{—}x\text{—} \equiv x = y)\ \&\ \ldots y \ldots)\text{'}$$

so class-descriptions may be introduced by an analogous contextual definition

$$\text{`}(\ldots\hat{x}(\text{—}x\text{—})\ldots)\text{'} \quad \text{for} \quad \text{`}(\exists y)((\forall x)(\text{—}x\text{—} \equiv x \in y)\ \&\ \ldots y \ldots)\text{'}$$

and it will then turn out that the 'paradoxes' we have been going through result quite simply in the conclusion that there is no such class as the Russell class was supposed to be. I take it that the advantages of this approach in the case of classes do not need arguing nowadays.[12]

In the case of the expression interpretation one is inclined to think that the first assumption is quite undeniable. For it is certainly very natural to suppose that the expression formed by taking a predicate-expression, enclosing it in quotation marks, and adding the prefix 'the expression . . .', *must* succeed in

[12] The point is argued explicitly by Goddard in 'Sense and Nonsense', *Mind*, 1964 (though he fails to do justice to the fact that Russell himself held a 'no-class' theory), and it is the main motivation for Quine's system NF. See *From a Logical Point of View*, chapter 5, pp. 91–3.

denoting an expression. But on reflection this thesis appears to me to be unconvincing, for our use of quotation marks is not nearly so pellucid as it suggests. First, it seems clear that quotation marks have no *one* clear-cut use but many different uses which may be indicated by the different prefixes that can be attached to them. For instance we can speak of *the word* 'now', *the series of letters* 'now', and *the pattern of marks* 'now', and in each case something different is being talked of. (For instance the word 'now' is the same word as the word 'noo' as that occurs in representations of Scottish speech, but the series of letters 'now' is not all the same thing as the series of letters 'noo'; again the series of letters 'now' is the same series of letters as the series of letters 'NOW', but this is not true of the pattern of marks 'now'.) It seems natural to explain the first two of these uses as truncated definite descriptions relying on the third, in such a way that

> the word 'now'

is construed as short for *something* like

> the word which, when written in standard English spelling, is written by writing the series of letters 'now' (and leaving a space before and after it)

while similarly

> the series of letters 'now'

may be construed as short for something like

> the series of letters which, when written in lower-case English script, appears as the pattern of marks 'now'.

It seems possible to construe nearly all ordinary uses of quotation marks in this fashion, as yielding descriptions which ultimately rely on a reference to the pattern of marks displayed within them, and if this is so it will certainly make sense to deny the existence of the word, or whatever it is, that is so described.

To illustrate, let us apply this view to the Russellian version of the paradox, where the relevant proposition to consider is

> For every expression x, the expression '. . . is not true of itself' is true of x if and only if x is not true of itself.

Now if the expression

> the expression '. . . is not true of itself'

does indeed succeed in referring to an expression, then the proposition we are considering is contradictory, but one is also very strongly inclined to say that if it does succeed in referring to an expression then, because of the nature of that expression, the proposition must on the contrary be true. The solution I suggest is that we should deny that there is any expression here referred to; there is no such thing as the expression '. . . is not true of itself'. I do not deny that there is such a thing as *the pattern of marks* '. . . is not true of itself', for such a denial would seem to be self-stultifying, but I take it that it cannot merely be the pattern of marks which is in question here, since truth can hardly be construed as a property of patterns of marks. The sort of thing that may be true of something is surely not a pattern of marks but something which stands in a certain relation to the pattern of marks, and what we can sensibly deny is that there is anything which stands in the required relation to this particular pattern of marks. (And this is not to deny our proposition a sense, but only to deny its ostensible referring-expression a reference.)

Now I would not care to argue that the way of construing quotation marks just suggested is the only possible way of construing them, for I am not at all convinced that if I were to write, say, of *the command* 'now' or *the reply* 'now' or indeed *the concept* 'now' I should have to be construed as referring ultimately to the pattern of marks 'now'.[13] In the last paragraph, for instance, I remarked that if the expression 'the expression ''. . . is not true of itself''' does indeed refer to an expression, then we seem *forced* to conclude that the paradoxical proposition is true because of the *nature* of the expression referred to. And this suggests the possibility that it may perhaps be our way of referring to the alleged expression that shows that it would have to have the nature in question; or in other words, the alleged expression may be being introduced not by specifying a pattern of marks to which it stands in a certain relation but more directly by specifying its truth-conditions. Roughly the idea is that expressions of the type

the expression '. . . is ϕ'

[13] See Goddard and Routley: 'Use, Mention, & Quotation', *Australasian Journal of Philosophy*, 1966. The arguments they employ do not seem to me *compelling*, at least until they raise the question of quantifying into quotation marks, but they are certainly *persuasive*.

should be construed as descriptions of the general sort[14]

> the expression which is true of a thing if and only if that thing is ϕ.

If they are construed in this way, then the expression

> the expression '. . . is not true of itself'

will be construed as the description

> the expression which is true of a thing if and only if that thing is not true of itself

and this would certainly account for our inclination to say that the paradoxical proposition *must* be true. But it is now quite obvious that there is no such thing as the expression '. . . is not true of itself', for it is an entirely straightforward matter to show that there is no expression which is true of every expression which is not true of itself. And, to revert to my earlier theme, it is of no use whatever to object that the expression 'heterological' is defined precisely as an expression which is true of every expression which is not true of itself, for definitions must obviously be incapable of creating what can be proved not to exist.

But enough of quotation marks. To sum up this section, the immediate consequence of Russell's paradox is that, wherever it seems plausible to say that a type of expression '$\hat{x}(\text{— x —})$' may be thought of as *mentioning* what is *used* of x when it is said that — x —, we cannot accept both of the assumptions

(i) that '$\hat{x}(Fx)$' always denotes something
(ii) that '$F(\hat{x}(Fx))$' always makes sense.

Yet both of these assumptions would seem to be intuitively acceptable (even though we might also want to add that '$F(\hat{x}(Fx))$' was very seldom *true*). My preference is for keeping assumption (ii) and rejecting assumption (i), even in the case of the expression-interpretation, because it seems to me less counter-intuitive to rule that certain prima facie truths are really false than it does to rule that they are really nonsense. But in a way it does not matter to my general argument precisely what solution of the problem is adopted, for the chief and most

[14] I am being deliberately vague here, because it may well be objected that a truth-functional 'if and only if' is not a strong enough connective in this context, and I do not want to embark on a discussion of what should be put in its place.

important moral to draw is that we cannot assume the principle of abstraction without more ado; we cannot assume that the switch from use to mention is always unproblematical, and any case in which we require to make this switch should be noticed explicitly as a new departure. With my decision always to treat (i) as the dubious assumption, rather than (ii), the 'new departure' figures as an assumption of *existence* rather than an assumption of *sense*, but—as I say—I do not think the opposite choice would make much difference to the remainder of my argument.

3. Subjects and Predicates

I have said that on any occasion on which we switch from using a predicate to mentioning that predicate an existential assumption is being made, and this is a genuine assumption which requires to be noted explicitly, for a situation can arise in which we have to recognize that the assumption is false. The moral then would seem to be that we should always try to avoid mentioning predicates, but—if we are not to give up logic entirely—this is liable to seem an intolerable requirement. Of course there is no difficulty *within* our formal system, for there it is easy enough to ensure that the principle of abstraction and similar principles do not figure among our fundamental rules and axioms, but there does seem to be a considerable difficulty in obeying this requirement when we are speaking *about* the formal system. For such discourse evidently has to take place in the common language we all use, and this language appears to commit us to the truth of the abstraction principle in all cases. It is after all just because we are ordinarily prepared to switch from use to mention without further ado that we find the falsehood of the abstraction principle so puzzling, and it is I think difficult not to sympathize with Russell's early insistence that there is some *contradiction* in the notion of a thing that cannot appear as a logical subject.[15] It does certainly seem that whatever the topic of our discourse is that topic will be indicated in our discourse by a grammatical subject-expression, and it will accordingly be hard to deny that it appears in our discourse as a logical subject.

But it is now time to reconsider the distinction just hinted at between a grammatical subject-expression and a logical subject.

[15] *Principles of Mathematics*, chapter 4, section 49.

I said earlier that we could not evade our problem by restricting logical subject-expressions to expressions which mention things of one or another special sort, for any expression used to mention anything gives rise to exactly the same logical problem. And it is for this reason that I take the notion of a logical subject to be fundamentally connected with the notion of mentioning or referring. But what we can do is to insist that several expressions which are grammatical subject-expressions do *not* really occur as expressions used to mention anything—or at least that they do not mention anything in the way in which a logical subject-expression would. It is after all the whole point of Russell's theory of definite descriptions that although a definite description is a grammatical subject-expression we do not *need* to take it as an expression for a logical subject, for we know how to paraphrase every proposition in which it occurs in such a way that this grammatical subject-expression disappears into a pattern of quantifiers and identity. In this case we can use what is grammatically a subject-expression while insisting that we are not introducing a logical subject precisely because the subject-expression can be *eliminated*, for from this it follows that everything that could be said with the subject-expression could equally well be said without it, and so no *new* assumptions are introduced by speaking in the one idiom rather than in the other.

Now this principle can be applied generally. Suppose, for instance, that we wished to talk of the concept *horse*, and for this purpose we employed, as seems natural, the grammatical subject-expression 'the concept *horse*'. Then if *all* we wanted to say about this concept were that everything that falls under it also falls under the concept *animal*, and if we are willing to accept that our remark could equally well be put in the form 'Whatever is a horse is an animal', I would say that we would not be committed to recognizing the expression 'the concept *horse*' as a logical subject-expression. Evidently the same holds for such similar locutions as

Being a horse materially implies being an animal,
The class of horses is included in the class of animals,
The expression '. . . is an animal' is true of whatever the expression '. . . is a horse' is true of,

and so on. Generally, if a certain grammatical subject-expression is used *only* in contexts where we can eliminate that subject-expression in the manner indicated, then there is no call to take the use of that expression as the use of a logical subject-expression.[16]

I say, then, that a proposition should be taken to have a logical subject if it is expressed with the help of a grammatical subject-expression which is not taken to be eliminable. The position is that use of a grammatical subject-expression is prima facie evidence that a logical subject is involved, but this evidence is 'defeasible'. One way of defeating it is by offering a paraphrase in which the grammatical subject-expression is eliminated in the manner just indicated, but I do not think that we need to stipulate that this is the only way of defeating it. For the essential feature of this elimination is that the grammatical subject-expression occurs only in a certain range of contexts which are to be understood *as wholes*, and though it is evidently highly satisfactory to be able to demonstrate this by actually paraphrasing the whole, it seems quite reasonable to allow that in some cases the common idiom might not in fact have the vocabulary required for a paraphrase. Thus one would probably agree with Quine[17] that expressions of the form 'for the sake of x' do not contain logical subject-expressions 'the sake of x'; the 'logical units' here are not expressed by 'for' and 'the sake of x' but rather by 'for the sake of' and 'x'. Now it would help to establish this point if one could offer a paraphrase, such as 'in order to benefit x', which eliminated the unwanted subject-expression, but it might turn out that no suitable vocabulary was available. (Evidently the paraphrase just suggested is in many ways inadequate.) In such a case one would simply have to proceed by showing how the phrase 'for the sake of' is in fact introduced as a unity, and how all other uses of 'sake' may easily be understood in terms of this one, for this would be to show that nothing would be lost by introducing a single word in place of the phrase 'for the sake of' and thereby eliminating the expressions 'the sake of x'. Accordingly I take it that it would be to show that these expressions are eliminable, as it were 'in principle'.

At this point I can explain what I mean by a *logical object*, for what I mean by it is pretty much what I mean by a *logical subject*.

[16] Compare the approach adopted in P. T. Geach, 'Class and Concept', *Philosophical Review*, 1955.

[17] *Word and Object*, p. 236.

I say, then, that a theory takes something to be a logical object if and only if either (i) grammatical subject-expressions apparently mentioning that thing are in the theory taken to be logical subject-expressions, or (ii) the subject-variables of that theory are taken as including that thing within their range. (And I shall have more to say about subject-variables in the next chapter.) But when I speak of a theory here I do not mean just the formal system in which that theory may be, as it were, crystallized, for given *only* the formal system one cannot even tell, for instance, whether it has any *subject*-variables at all. I mean, rather, the formal system together with its interpretation, or in other words together with all those propositions which we need to be given if we are to understand what the formulae of the system mean. And this brings me back to the difficulty I raised at the beginning of this section: obviously we do not want it to turn out that every theory which has anything to say about predicates is committed to treating predicates as logical objects, and yet every theory which has anything to say about predicates must surely explain what predicates are, and how could one possibly explain what predicates are without mentioning predicates? Since this problem affords a useful paradigm for the case of numbers, I will now consider it more closely.

The formal system of ordinary first-order predicate calculus does not contain any expression which seems to mention a predicate or predicates, since predicate-expressions and schematic predicate-letters occur in that symbolism only in immediate conjunction with subject-letters or subject-variables, which shows that they are to be taken as expressions or schematic letters for predicates in use and not for mentionings of predicates. The problem, however, is to explain what this symbolism *means* without at any point mentioning predicates. Now the symbolism employs formulae such as 'Fa', 'Fa ∨ Ga', '(∀x)(Fx ∨ Gx)', and —as will appear more fully in the next chapter—these occur only in the context 'proposition of the form . . .'. The main task, then, is to explain the meaning of such phrases as 'proposition of the form "Fa"', 'proposition of the form "Fa ∨ Ga"', 'proposition of the form "(∀x)(Fx ∨ Gx)"', and so on.

It would seem that we ought to proceed here by first explaining the notion of a proposition and then explaining the notion of a propositional form, but the first of these tasks is one that I shall not attempt here. I shall here presuppose the notion of a proposition, and in what follows I shall preserve a distinction between a proposition and a propositional expression (treating the latter as if it were a sentence). Similarly I distinguish between a subject and a subject-expression, and between a predicate and a predicate-expression, saying that propositions contain subjects and predicates while propositional expressions contain subject-expressions and predicate-expressions. In making this distinction I intend to allow for those analyses of propositions which distinguish propositions from the expressions for them, while not necessarily insisting that such a distinction must be drawn, for my intention is to remain neutral between all accounts of propositions which satisfy certain minimum requirements. There are obviously various rather different accounts of propositions, which we may roughly classify as those that start from the notion of a sentence, those that start from the notion of a speech-act, and those that start from the notion of a thought, belief, judgement, etc.; but so long as some account can be given of what is to count as a proposition and of what is to count as the same proposition, it does not matter for my present purpose what exactly that account is. What I require of any account of propositions is just that propositions should turn out to have the appropriate connection with truth and falsity. Thus it must at least be the case that every proposition is either true or false and that no proposition is both true and false. Furthermore I assume, at least for my present purpose, that we may safely adopt an abstraction principle for propositions and truth, namely that for all propositions p there is something, viz: the proposition that p, that is true if and only if p.[18] It may perhaps turn out that this assumption proves, on closer examination, to be unnecessary, but since it is predicates and not propositions that have been causing our difficulties so far I shall here simply assume it, so that I can mention and refer to propositions freely in what follows.

[18] The principle as stated here has the appearance of a tautology, but this appearance is misleading, as will emerge more clearly from my discussion of quantification in the next chapter. The principle gives rise to no Russell-like contradiction, so far as I know, provided that we refuse to recognize as a proposition anything which would—if it were a proposition—be a proposition about its own truth-value. (Though this proviso needs to be stated more carefully.)

Turning, then, to consider propositional *forms*, our intuitive notion here seems most naturally explained in connection with the idea of analysing propositions as 'containing', or as 'constructed from', certain *components*—for instance other propositions, or (as we commonly suppose) subjects, predicates, and so on. For to give such an analysis of a proposition will be first to specify the components it contains and then to explain *in what way* it contains them, i.e. to explain how the proposition is constructed out of those components. But this latter account would usually be given in general terms so as to be applicable also to other propositions constructed in the same way out of other components, and the one way in which these various propositions were constructed out of their various components could reasonably be taken to be a propositional form. This notion of form or structure is of course relative to the notion of a component in the sense that different accounts of the structure of a proposition will be required for different accounts of its components, and there is no need to suppose that any one account of the components of a proposition must be the final account; in other words analysis may always be pushed deeper and deeper—or at least there is no reason to suppose that it may not. It may be tempting to say that whatever level of analysis has been reached the structure can be regarded as a kind of skeleton which is what remains of the proposition once all components have been subtracted, but I think that this analogy is misleading. For in the limiting case when the proposition is taken to have itself as sole component it seems clear that subtraction of the component would leave nothing remaining at all, so that in this case there would be no structure to be explained. But in fact there is still a way in which this component is used to form the original proposition, just as there is a (different) way in which this same component could be used to form the negation of the original proposition; to list something as a component is not to say in what way that component is to be used to construct the proposition desired.

A propositional form, then, one might perhaps think of as a structure of propositional components, of such a sort that to specify a propositional form is to give a rule or set of directions for constructing a proposition from any components of certain types; or one might perhaps think of it as a correlation or mapping which takes us from an ordered set of propositional components to a proposition. Thus the form 'P & Q', for instance, might be taken to be the correlation that maps any ordered pair of propositions onto the proposition that is the conjunction of the first with the second, and the form '∼P' will similarly be the correlation that maps any proposition onto the proposition that is its negation, the form 'P' will be the correlation that maps any proposition onto itself, and so on. Generally, the number and type of the components that are mapped by some form onto propositions will be indicated by the number and type of the schematic letters occurring in the formula by which it is specified, and the way in which those components are to be constructed into a proposition will be indicated by the way in which those schematic letters are arranged, with the other signs, in the formula.

Now if we are going to produce an answer to the question 'What is a propositional form?' the sort of answer just sketched seems to me a natural one to give. But it should be noticed that it is quite possible to avoid this question altogether. For we shall want to make use of the notion of a propositional form only when we want to speak of propositions being *of* that form, and for this purpose there is no need to specify what is meant by the words 'the form . . .'; it will be quite sufficient if we explain the whole phrase '. . . is of the form . . .' taken as indissoluble. The way in which this whole phrase is to be explained is easily enough seen from my previous examples. For instance, instead of saying that *the form* 'P ∨ ∼P' is the correlation that maps any proposition onto the proposition formed by disjoining it with its own negation, we may merely rephrase the explanation so that it states that a proposition is *of the form* 'P ∨ ∼P' if and only if it can be analysed as containing one proposition as component and as constructed from that component by disjoining it with its own negation. Provided that the expression 'the form "P ∨ ∼P"' occurs only in the context '. . . is of the form "P ∨ ∼P"', the two explanations clearly come to exactly the same.

I shall not delay here to construct a general explanation on these lines for all formulae of the propositional calculus, for this is somewhat tedious, but it should be clear from my examples how it is to be done. It is clear too that we may easily extend the explanation to any formula in which actual propositional expressions are substituted for the schematic letters in these formulae, so that we can speak of a proposition as being of the form, say, 'P & the earth is flat'. (This means evidently that the proposition can be analysed as containing one proposition as component and as constructed from that component by conjoining it with the proposition here expressed by 'the earth is flat'.) In fact it proves technically convenient to extend this idea so far as to allow that a formula may be used to represent a propositional form even when it contains no schematic letters at all, so that we may speak, for instance, of a proposition as of the form 'the earth is flat', meaning by this that it *is* the proposition expressed by 'the earth is flat', and I shall make some use of this stipulation later.[19] Anyway there is, as I say, no difficulty in giving this type of explanation of formulae involving schematic letters for propositions.

But of course it is clear that the reason why I can maintain that there is no difficulty in explaining expressions such as 'proposition of the form "P & Q"' is that in my explanation I have been freely speaking of propositions in a way that would apparently commit me to recognizing propositions as objects (and it was precisely to legitimize this that I began by assuming an abstraction principle for propositions). We cannot simply transfer the explanation to expressions such as 'proposition of the form "Fa & Ga"' without speaking in a way that apparently commits us to recognizing as objects the relevant components in this case too, namely subjects and predicates; but this last is exactly what we wish to avoid. There is, however, one very relevant difference between propositions and the various further components of propositions that we may wish to recognize, and that is that unlike propositions themselves these further components will always occur in propositions and never in isolation. Consequently it should always be possible to say what we need to say without actually referring to the components but instead referring to the propositions that contain them. Roughly speaking, whatever type of propositional components α-components may be, the important notions will be the notions of a proposition *containing an* α-*component* and the notion of several propositions *containing the same* α-*component*, and once these are

[19] See chapter 3, text to notes 4 and 7. This convention would be a natural consequence of the view that propositional forms are rules for constructing propositions out of certain sets of components, for what we have here is a rule for constructing a proposition out of the null set of components.

understood we shall have all we need to explain the use of schematic α-letters in representing propositional forms. Our problem, therefore, is just to give an explanation of these two locutions *as wholes.*

More precisely, in order to introduce the symbolism of the monadic predicate calculus, we would begin by introducing the two series of letters 'a', 'b', 'c',..., and 'F', 'G', 'H',..., and calling these the series of schematic subject-letters and predicate-letters respectively. From these we construct the elementary subject-predicate formulae which consist of a predicate-letter followed immediately by a subject-letter, and we explain what it means to say that a proposition is of such an elementary form, e.g. that it is of the form 'Fa'. The explanation here is quite simply that a proposition is of such a form if and only if it *contains a subject*, and in such a case I shall also say that it *contains a predicate occurring initially*. (By a 'predicate occurring initially' I understand intuitively what is left of a proposition when some subject it contains is dropped out, but I do not offer this as a definition; on the contrary all that I define is the whole phrase 'contains a predicate occurring initially', and this is defined as a mere variant on 'contains a subject'.) Now we also have to explain the significance of a formula which contains repetitions of the same subject-letter or the same predicate-letter, as for instance 'Fa & Ga' or 'Fa & Fb', and in explanation of this point we may say (i) that a proposition is of the form 'Fa & Ga' if and only if it may be analysed as containing two propositions as components, each of which *contains the same subject*, and as constructed from those components by conjoining them; while (ii) a proposition is of the form 'Fa & Fb' if and only if it may be analysed as containing two propositions as components, each of which *contains the same predicate occurring initially*, and as constructed from those components again by conjoining them. If we say generally that wherever we have several formulae containing the same subject-letter, as 'Fa', 'Ga', 'Ha',..., then several propositions may be said to be of these *related* forms if and only if they each contain the same subject, and similarly for predicates, then in terms of this it will be a simple matter to explain the use of any formula which is a truthfunctional combination of elementary subject-predicate formulae (and perhaps proposition-letters). What remains to be given is the explanation of formulae involving quantifiers, and that is a topic I shall consider in the next chapter.

We see, then, that in the case of the monadic predicate calculus the three basic notions that need to be explained are the notions of containing a subject, containing the same subject, and containing the same predicate initially. I have already indicated how the first of these is to be understood, and I do not intend to elaborate on this further except by pointing out that what I have offered is an explanation of the *whole* phrase 'contains a subject'. I have not offered any account of *what a subject is*—whether for instance it is the thing referred to, the expression which refers to it, the act of referring to it, the concept under which the thing is being thought of, or what—and I see no necessity to offer any such account. As regards the notions of containing the same subject and containing the same predicate initially, it should be observed that these are relations between propositions and my point is again simply that they should be explained directly as such. The relations are symmetrical and under certain restrictions transitive,[20] and Russell has often observed that we tend to construe symmetrical and transitive relations as signifying the presence of some third thing common to the two terms of the relation. Thus the relation '. . . is as high as . . .' is a symmetrical and transitive relation, and we tend to think that if x is as high as y this must be because there is some third thing, a height, such that x has that height and y has that height. But if this version of the principle of abstraction is assumed unrestrictedly for all symmetrical and transitive relations, then we have seen that it leads at once to Russell's paradox, and besides it is not clear to me why the principle should be thought to have any explanatory value at all. In fact it is altogether more plausible to suppose that the relation '. . . is as high as . . .' is to be understood *first*, and then in the light of this we may perhaps come to construct the further notion of a height if there seems any reason to do so. Similarly I suggest that we should first impart an understanding of the whole relations '. . . contains the same subject as . . .' and '. . . contains the same

[20] The relations are transitive within the field of propositions which contain just one subject or predicate. And a proposition contains just one subject (or predicate) if and only if all propositions which contain the same subject (or predicate) as it also contain the same subject (or predicate) as each other.

predicate initially as . . .' without drawing upon such alleged entities as subjects and predicates. For instance there is no need to decide what sort of thing a predicate is, nor therefore what is to count as *being* the same predicate, in order to introduce the relation which holds between two propositions when they *contain* the same predicate (initially). And, as I have been saying, it is only this latter relation, and the analogous relation for subjects, that needs to be explained if we are to introduce the formulae of the monadic predicate calculus.

As for the question of how these relations are to be explained, it seems to me that the procedure here must be largely ostensive, i.e. it must rely to a large extent on our simply citing examples of pairs of propositions between which the relations do or do not hold. We do not have to come to this stage at once, for it is entirely accurate to say that two propositions contain the same predicate initially if and only if it is possible to obtain the same proposition from each by substituting the same subject for some subject of each; but now we have to explain the relevant operation of substitution, and since there seems no plausible escape from ostension here it is not clear that the gain is appreciable. Naturally the ostension may be controlled by stating some of the conditions which these relations must satisfy—for instance any proposition which contains a subject contains the same subject as itself, and if two propositions contain the same subject then so do any other two propositions which contain these propositions as components, and so on—and we may also try to help by drawing various sorts of analogy and using various metaphors, though these must in their turn be understood in the light of the literal examples given. But it is no objection to my position that we are reduced to examples at this point, so that from a formal point of view these relations are taken as primitive, for it is obvious that any chain of explanations will come to an end in examples somewhere, and I do not see why it is unreasonable to end where I suggest.

Now I should certainly admit that the monadic predicates so far considered are a particularly simple case, and we should need to introduce further complications if we were dealing with polyadic predicates generally, but the principle upon which I have been proceeding is clearly applicable to any such case. In each case we begin by framing our explanations of the relevant formulae by speaking of propositions containing the appropriate components in various ways, and we then introduce these further notions as wholes without referring to the components in question. Indeed the general possibility of the 'reductive' account, for so far as we have taken it, may perhaps seem quite unsurprising, for I think it is generally held that it is not the use of schematic letters for predicates or other components that commits us to recognizing them as objects, but rather the use of quantified variables. This brings me to the subject of my next chapter, but before I proceed to this it may be helpful if I here quickly recapitulate the main outline of my argument so far.

I began the chapter by rejecting Russell's vicious-circle principle. The discussion here introduced several points about definitions which will be pursued further in chapter 6, section 2, but its immediate importance is largely negative: the *ramified* theory of types should be rejected, and the *simple* theory cannot be developed from Russell's basis. Nevertheless an adequate basis for a theory of types can be found in Russell's discovery that the principle of abstraction is self-contradictory, for this discovery points up the distinction between *using* a predicate and *mentioning* a subject in a very sharp way. To put the matter succinctly, it cannot be assumed that whatever can be used can be mentioned, though we can certainly construct expressions which *purport* to mention the thing, just as a definite description which in fact fails of reference will nevertheless *purport* to refer. Hence we have good ground for rejecting the form 'F(F)' and insisting that what is intended be written differently, say as '$F(\hat{x}(Fx))$', where we recognize that the part '$\hat{x}(Fx)$' cannot be guaranteed a reference just on the ground that 'F ...' has a sense. In fact we should restrict our schematic predicate letters to contexts where they supplant predicates in use, and these are contexts where the predicate-expression occurs in immediate conjunction with an expression purporting to mention something, or with a quantified variable which takes the place of such referring expressions.

Now there are several expressions which have the grammatical features of expressions purporting to mention something, but which can be 'eliminated' in the sense that the propositions they are used to express are agreed to be expressed equally well by an alternative form of words containing no such expression. The occurrence of such expressions should evidently be viewed as a mere matter of idiom, which may be expected to vary from one language to another, and is clearly of no logical importance. When these expressions are discounted, the propositions which still require expressions purporting to mention something comprise an important logical category, for we have seen that all of these propositions, and only these propositions together with such as are got from them by generalization, will contain a predicate in use. Hence I say that it is just these propositions which contain a subject, and this gives the rationale for the principle which I say lies at the heart of the theory of types, viz. that a predicate-letter has a use only in immediate conjunction with a subject-letter or a subject-variable. Further, I say that a theory takes something as a 'logical object' if either the theory contains non-eliminable subject-expressions purporting to mention the thing, or it employs subject-variables which are intended to include that thing within their range, and this is why I remarked earlier that the thesis that numbers are not objects is effectively the thesis that the apparent referring role of numerals is eliminable.

Finally, when we consider the interpretation of logical formulae in the light of this discussion, it seems difficult—though perhaps not impossible—to avoid treating *propositions* as logical objects, since they constitute the subject-matter of logic itself. But there need be no reason to treat other propositional components in the same way. At least, what I have argued for so far is that one can introduce *schematic letters* for other propositional components without at any point treating those components as logical objects; it remains to be seen whether the same holds good for quantified variables.

3. Quantification

1. The Analysis of Quantification

In logic we analyse propositions as being of this or that form because we very often wish to speak of *every* proposition of a certain form, saying perhaps that every such proposition is true, that not every such proposition is false, and so on. Provided we understand what it is for something to be a proposition, for a proposition to be true (or false), and for a proposition to be of the form in question, then I do not see that we need to give any further philosophical explanation before we can speak of *every* proposition of that form being true; in other words, the notion of generality which enters here seems to me incapable of useful philosophical elucidation. But we shall need to introduce a further refinement before we can use this notion of generality to its fullest effect.

To make my discussion easier to follow here I shall begin by referring freely to propositional forms and propositional components, saying that a form 'calls for' components of various types according as the relevant formula contains schematic letters of those types. (Thus where a formula contains, say, two schematic subject-letters and one schematic predicate-letter, I shall say that it represents a form which 'calls for' two subject-components and one predicate-component, the idea being that when these are 'supplied' the form will 'construct' a proposition from them.) At the end of the chapter I shall show how this extravagant language can be eliminated, but I think it will help if I make free use of it to begin with.

Now I say that in order to make full use of the notion of generality we need to introduce a further refinement, and this need may be brought out by considering propositions of a form which calls for two or more components (of the same or of different types). For example suppose we start with some proposition which could be expressed by

John loves Mary

which we will assume to be a subject-predicate proposition analysable indifferently as containing a subject expressed by 'John' or as containing a subject expressed by 'Mary'. Then the proposition may be said to be of the form

a loves Mary,

and we may have occasion to consider the proposition

Every proposition of the form 'a loves Mary' is true.

But now this *general* proposition may itself be analysed as a subject-predicate proposition, having a subject-component (expressed by 'Mary') from which the proposition is constructed by a certain form. However, it is clear that we cannot represent that form in the usual way, namely by writing the above sentence with another schematic subject-letter in the place of 'Mary' so as to get

Every proposition of the form 'a loves b' is true,

for *that* expression is still a complete sentence expressing a complete proposition, but what we wanted was a formula to represent a propositional form that called for a further subject-component.

Evidently the trouble is that *both* the schematic letters in 'a loves b' are captured by the prefix 'the form', whereas we intended only the first to be. We must, then, introduce some device to show that the newly introduced schematic letter 'b' has nothing to do with the old prefix. We might try to achieve this by adopting the convention that where an expression for a propositional form occurs within the context

Every proposition of the form . . . is true

we write the schematic letters of that form differently, say by using 'x', 'y', 'z',... in place of 'a', 'b', 'c',.... In that case the proposition we were considering should be written rather as

Every proposition of the form 'x loves Mary' is true

and we can now go on to say that this general proposition is itself of the form

Every proposition of the form 'x loves b' is true,

where the difference in alphabet between 'x' and 'b' shows that it is only the former that is in the relevant sense *within* the context 'Every proposition of the form . . . is true'.

But of course this has only postponed the difficulty, for we may want to generalize our proposition still further, so as to say, for instance

Not every proposition of the form 'Every proposition of the form "x loves b" is true' is true

and by our previous ruling it seems that the letter 'b' must in this context be exchanged in its turn for a letter of the new alphabet, say 'y', and so we are back to our initial dilemma. And if we introduce yet another alphabet to deal with these double prefixes, then evidently we shall have to introduce still further alphabets to deal with still more complicated prefixes, and the whole system threatens to become unworkable. Clearly what is wanted is some more direct way of *linking* a prefix with the schematic letters with which it is concerned, which one might achieve by drawing appropriate arrows thus:[1]

Not every proposition of the form
every proposition of the form 'a loves b' is true'
is true.

But this notation again proves cumbersome in more complicated cases, and the most convenient plan is certainly to repeat the relevant schematic letters (or variables) within the prefix. Thus instead of

the form 'a loves Mary'

we may also when necessary write

the form *with respect to* 'a' 'a loves Mary'

or more briefly

the form$_{(a)}$ 'a loves Mary'.

With this convention we may without ambiguity put our doubly general proposition by saying

Not every proposition of the form $_{(b)}$ 'Every proposition of the form$_{(a)}$ "a loves b" is true' is true

and it is clear that the technique is easily extended to yet more complicated cases.

Now it is time to introduce some abbreviations, and in place of

Every proposition of the form '(——)' is true

[1] Compare Quine, *Mathematical Logic*, § 12.

I now propose to write

$$\vdash(\text{——})$$

while in place of

Every proposition of the form$_{(a)}$ '(—a—)' is true

I now propose to write

$$(\forall x)(\text{—}x\text{—})$$

and similarly for any other schematic letter.[2] But it should be evident that I do not mean this to be a mere abbreviation, but rather an analysis of the universal quantifier and of the significance of something's being said to be a logical theorem or axiom. And this proposal demands some discussion.

Of course when it is said that some formula is a theorem all that need strictly be *meant* is that either it is one of the formulae given as 'initial formulae' of the system in question, or it can be obtained from these formulae by certain operations given as the 'transformation rules' of the system, but that is not what I have in mind when I speak of the *significance* of something's being a theorem or axiom. To understand this we have to take into account the intended interpretation of our system and to adopt the attitude in which it makes sense to ask whether something is *acceptable* as an axiom, whether something *deserves* to be a theorem, and so on. Roughly speaking, we must adopt an attitude in which laying down some formula as an 'initial formula' is seen as *assuming* that something is *true*, and when I ask for the significance of something's being taken as an axiom I am asking what it is that is thereby being assumed to be true. (And the significance of something's being a theorem is to be understood analogously.)

Now in nearly all cases it will be clear that a formula of a logical system is to be taken as representing a propositional form, and to assume some formula as an axiom or to prove that it is a theorem is to assume or prove something of the propositional form in question, namely that all propositions of that form are true. More usually it is said that what is assumed or proved is that 'the formula comes out true for all interpretations of its schematic letters', or something else of this sort, but I think that this differs only verbally from what I have been suggesting, and

[2] These abbreviating conventions will be explained more exactly in the next section.

I regard this part of my proposal as relatively uncontroversial. An objection that might be raised is that being a logical axiom or theorem ought not to be taken to be a matter of mere *de facto* universality, for in logic we are concerned to state not merely that all propositions of certain forms are true but further that they are in some sense *necessarily* true. I am not convinced that this point need in fact conflict with the suggestion I am making, for it would seem possible to insist upon two stages here, maintaining (i) that what is signified by writing, say, 'P v ∼P' under the heading 'theorems' is what I say, i.e. it is just that all propositions of the form 'P v ∼P' are true; but adding (ii) that what is meant by saying that the following theorems are theorems *of logic* (as opposed to e.g. theorems of physics) is the further point that the propositions expressed by writing these formulae as theorems are not only true but also necessarily true. If, however, this distinction seems specious I would in fact be quite prepared to grant the objection since it does not affect my main line of argument otherwise than by introducing a tedious complication. All that I need to be granted is that to assume or prove that some formula is a theorem is *at least* to assume or prove that all propositions of that form are true, and it will not matter if we go on to add that it is also to assume or prove something stronger than this.[3]

As regards the point that I need to be granted, one could perhaps support this by considering the rule of uniform substitution for schematic letters, which holds for schematic letters of any type in all standard logical systems. For instance from the fact that 'P v ∼P' is a theorem we are entitled to infer that '(P & Q) v ∼(P & Q)' is a theorem, that 'Fa v ∼Fa' is a theorem, that '(∀x)(Fx) v ∼(∀x)(Fx)' is a theorem, and so on. If one were asked to justify this principle of inference it seems to me that one would very naturally draw attention to the fact that any propositions of the form '(∀x)(Fx)' or of the form 'Fa' or of the form 'P & Q' are also of the form 'P', so that any propositions of the form '(∀x)(Fx) v ∼(∀x)(Fx)' or of the form 'Fa v ∼Fa' or of the form '(P & Q) v ∼(P & Q)' must also be of the form 'P v ∼P'. It then clearly follows that if every proposition of the form 'P v ∼P' is true, so also is every proposition of the form '(P & Q) v ∼(P & Q)' and so also is every proposition of

[3] See note 5.

the form 'Fa v ∼Fa' and so also is every proposition of the form '(∀x)(Fx) v ∼(∀x)(Fx)'; and this provides a justification for the rule of substitution if theoremhood is understood as I suggest. (And one can add the appropriate modal operators to this argument if necessary.) It is however not clear to me how the rule of substitution is to be justified if the account of theoremhood is importantly different from what I suggest.

Anyway, this part of my proposal is, I think, relatively uncontroversial. What is more controversial is the thesis that the significance of universal quantification is to be explained in the same way, but in fact if we accept the above account of what it is to be a theorem this thesis is pretty directly implied by the rules and axioms standardly adopted in connection with perfectly ordinary first-order quantification, at least so far as the first-order quantifiers are concerned. For, where 'a' is a schematic subject-letter and 'x' a variable for subject-components, ordinary first-order quantification theory commits us to the principle

'(—a—)' is a theorem if and only if '(∀x)(—x—)' is a theorem

and, granting my account of what it is to be a theorem this shows that a universal quantifier may be introduced when and only when we are speaking of all propositions of a certain form. The point emerges most clearly if we consider just the simple case in which the formula '(—a—)' contains no schematic letter other than 'a'; let it be, for instance, 'a = a'. Now to say that 'a = a' is a theorem is, if my first contention is correct, just to say that every proposition of the form 'a = a' is true, and similarly to say that '(∀x)(x = x)' is a theorem is to say that every proposition of the form '(∀x)(x = x)' is true. But since the formula '(∀x)(x = x)' is itself a complete sentence, expressing a complete proposition, there is only one proposition of this form, namely the proposition '(∀x)(x = x)',[4] and so in this case the principle reduces simply to

> Every proposition of the form 'a = a' is true if and only if the proposition '(∀x)(x = x)' is true

or in other words

> Every proposition of the form 'a = a' is true if and only if (∀x)(x = x).

[4] See chapter 2, text to note 19.

This principle is, I say, implied by perfectly orthodox rules of quantification theory together with what I take to be a fairly uncontroversial account of what is meant by saying that a formula is a theorem,[5] and it shows that in this simple case at least my account of quantification must be entirely adequate. And any other account, which does not at least imply mine, must either leave orthodox quantification theory without foundation or involve an unusual understanding of what it is to be a theorem.

In this simple case, then, where a formula contains only one schematic letter, the adequacy of my account of quantification may be established by appealing to the rules and axioms ordinarily adopted in orthodox quantification theory. However, one cannot produce a compelling argument on the same lines to show that the account is adequate in the more general case where a formula contains several schematic letters. For instance if we consider the formula 'Fa v ∼Fa' then the most that we can hope to establish by appealing to the rules and axioms of quantification theory is the interchangeability of the two propositions

> Every proposition of the form 'Every proposition of the form "Fa v ∼Fa" is true' is true (a)
>
> Every proposition of the form '(∀x)(Fx v ∼Fx)' is true

But this shows only that the two expressions

> Proposition of the form 'Every proposition of the form "Fa v ∼Fa" is true' (a)
>
> Proposition of the form '(∀x)(Fx v ∼Fx)'

are interchangeable in the special context

Every . . . is true

and it *might* be held that although they are interchangeable in this context they are not interchangeable in other relevant contexts. However, I do not see what ground one might have for holding this, and I think that it should at least be admitted that there is a plausible case here for the strong conclusion that the account of quantification I have been giving is *the* account which is required by orthodox quantification theory. As I shall presently

[5] If the objection noticed in the text to note 3 is accepted, then we should write instead: 'Every proposition of the form 'a = a' is *necessarily* true if and only if it is *necessarily* true that (∀x)(x = x)'. But it is clear that this modification would be entirely in keeping with my main argument.

demonstrate it is certainly *an* account which satisfies the requirements of orthodox quantification theory, and that is perhaps all that I need to claim for it.

The main conclusion that I want to draw from the account is that there is no distinction of philosophical importance to be drawn between the logician's use of schematic letters and his use of quantified variables. Of course for a concise statement of the rules and axioms of quantification theory it may well prove more convenient to employ two different alphabets, one for schematic letters and one for variables—indeed it has been suggested that pedagogical purposes are best served by using *three* such distinct alphabets[6]—but this distinction is only a matter of notational convenience; what is being said by means of the notation is in each case essentially the same. Naturally I do not mean that the symbol 'Fa v ∼Fa' is *always* interchangeable with the symbol '(∀x)(Fx v ∼Fx)', for it is the use of these symbols *to state theorems* that I am concerned with. And I do not mean that we must be saying *exactly* the same thing whether we write

'(Fa v ∼Fa)' is a theorem

or

'(∀x)(Fx v ∼Fx)' is a theorem.

The difference between them emerges in my paraphrase as the difference between

Every proposition of the form '(Fa v ∼Fa)' is true

and

Every proposition of the form 'Every proposition of the form$_{(a)}$ "(Fa v ∼Fa)' is true" is true.

And one might well say that the second of these is a more complicated statement than the first, for it does require for its expression both a double use of the prefix 'Every proposition of the form' and the auxiliary device for specifying which of the following schematic letters are to be regarded as bound by the prefix 'the form'. But it seems to me quite clear that these extra complications are the result simply of a deeper analysis of what is asserted by the first statement, and in particular they do *not* involve us in any further 'ontological commitment', nor are they

[6] See E. J. Lemmon, *Beginning Logic*, p. ix.

available only where it is subject-letters we are dealing with. With equal justice we could move from the first statement to

> Every proposition of the form 'Every proposition of the form$_{(F)}$ "(Fa v ∼Fa)" is true' is true,

i.e. to

> '(∀𝔣)(𝔣a v ∼𝔣a)' is a theorem.

All schematic letters here are on the same footing—even, one may add, the proposition-letters which are introduced at the initial stage of propositional calculus. If it is legitimate to use such letters to indicate generality in the expression of theorems, then it is equally legitimate to replace them by quantified variables, and the usual theorems of quantification theory hold equally for all such variables.

2. The Logic of Quantification

This last statement requires some elaboration, for so far I have argued only that the *meaning* of quantification is to be explained in the same way no matter what type of propositional component is in question, but I have not said anything about establishing the *laws* of quantification, and in fact it is not completely accurate to say that the laws of quantification are exactly the same for every type of propositional component. To see the situation more clearly here it is helpful to consider first a quite general theory of quantification, which we may present as the theory of quantification over the α-components of propositions, whatever α-components are taken to be, and then to inquire how this theory needs to be specialised for the various different propositional components.

The relevant laws here can be stated and justified more compendiously if we do temporarily disregard the notational distinction between schematic letters and variables, and I shall accordingly abandon it for the purposes of this section. I shall use the letter 'α' as a dummy schematic letter for α-components, whatever α-components are taken to be, and wherever it is necessary to mark the distinction between schematic letters proper and variables I shall instead speak of 'α' occurring free or bound in the relevant formula. More precisely, by a 'formula' I shall mean any series of signs which represents a propositional form,[7] and I use the signs '(—)' and '(...)' as metalogical schematic signs for formulae in general, while I use '(— α —)' and '(... α ...)' similarly for formulae containing some free occurrences of the letter 'α', i.e. containing the letter 'α' in some places where it functions *as* a schematic letter and not as a quantified variable. In other words the signs '(— α —)' and '(... α ...)' are schematic signs for formulae representing propositional forms which call for at least one α-component, and that they call for this α-component is here represented by the occurrence in them of the letter 'α'. The sign '⊢' may be prefixed to any formula to express the proposition that every proposition of the form represented by that formula is true, while the sign '(∀α)' may be prefixed to any formula containing 'α' free and is to mean the same as the words 'Every proposition of the form (with respect to the letter 'α') . . . is true'. (And in any formula '(∀α)(— α —)' with this prefix I shall say that 'α' occurs bound and not free.) Thus where some formula '(— α —)' has 'α' as its only schematic letter, a formula such as '(∀α)(— α —)' will express a complete proposition and be interchangeable with the formula '⊢(— α —)'; however, where '(— α —)' is taken as having further schematic letters the formula '(∀α)(— α —)' will represent a propositional form which calls for components to take the place of its other schematic letters (though not of 'α'), as was illustrated on pp. 56–8. The rest of my notation is self-explanatory.

Now the rules and axioms normally stated for quantification theory either are or are equivalent to the three principles

(i) If ⊢(— α —), then ⊢(∀α)(— α —) (Generalization)

(ii) ⊢((∀α)(— α —) ⊃ (— α —)) (Instantiation)

(iii) ⊢((∀α)(P ⊃ (— α —)) ⊃ (P ⊃ (∀α)(— α —))) (Distribution)

together with some slight modification to allow for the inter-substitutibility of different letters from the same alphabet. And it is not difficult to show by informal arguments that these three principles are available whatever components we take the α-components to be. First let us notice two simple consequences of our explanations of the notion of a propositional form, namely

(*a*) A proposition is of the form '(—)' if and only if a formula expressing that proposition can be obtained from the

[7] Note that formulae thus include complete sentences, by the stipulation in chapter 2, section 3, text to note 19.

formula '(—)' by substituting constants for any schematic letters there may be in that formula

(*b*) If the formula '(...)' can be obtained from the formula '(—)' by substituting constants for some of the schematic letters in that formula, then every proposition of the form '(...)' is also of the form '(—)', and so in particular if ⊢(—) then ⊢(...).

Now with these simple consequences in mind, we may easily establish the three principles stated above by a *reductio ad absurdum* technique.

(i) *Generalization.* First, if some proposition of the form '(∀α)(—α—)' is false, then by (*a*) that proposition must be obtainable from the formula '(∀α)(—α—)' by substituting constants for any schematic letters other than 'α' that there may be in '(—α—)', and so this false proposition is expressed by some formula '(∀α)(...α...)' where '(...α...)' is obtained by substitution of constants from '(—α—)'. But second, if the formula '(...α...)' is obtained by substitution of constants from '(—α—)', then by (*b*) we know that if ⊢(—α—) then ⊢(...α...). And third, we have already seen that in the case where '(...α...)' contains no schematic letters other than 'α', which is our present case, then '⊢(...α...)' and '(∀α)(...α...)' are interchangeable, so that if '(∀α)(...α...)' expresses a falsehood, then so does '⊢(...α...)', and it is not the case that ⊢(...α...). Putting these points together we deduce that if some proposition of the form '(∀α)(—α—)' is false, then it is not the case that ⊢(—α—), or in other words that if ⊢(—α—) then ⊢(∀α)(—α—).

(ii) *Instantiation.* If some proposition of the form '(∀α)(—α—) ⊃ (—α—)' is false, then by (*a*) this proposition must be obtainable by substitution of constants for the schematic letters in '(∀α)(—α—) ⊃ (—α—)', and hence its antecedent is obtainable by substitution of constants for the schematic letters (other than 'α') in '(∀α)(—α—)', and its consequent is obtainable by those same substitutions for the schematic letters other than 'α' in '(—α—)' together with a further substitution for 'α'. The proposition, then, will be expressed by some formula of the form '(∀α)(...α...) ⊃ (...)' where '(...)' contains no schematic letters and is of the form '(...α...)', and '(...α...)' contains no schematic letters other than 'α'. But in that case the antecedent states that every proposition of the form '$(\ldots\alpha\ldots)$' is true, and the consequent is indeed of that form and so must be true if the antecedent is true. It follows that the whole proposition '$(\forall\alpha)(\ldots\alpha\ldots) \supset (\ldots)$' must be true, contrary to hypothesis, and so we may conclude that no proposition of the form

$$`(\forall\alpha)(-\alpha-) \supset (-\alpha-)\text{'}$$

is false, or in other words $\vdash((\forall\alpha)(-\alpha-) \supset (-\alpha-))$.

(iii) *Distribution.* Again, if some proposition of the form '$(\forall\alpha)(P \supset (-\alpha-)) \supset (P \supset (\forall\alpha)(-\alpha-))$' is false, then such a proposition must be obtainable by substitution of constants in this formula, and so be expressible by some formula

$$`(\forall\alpha)((\ldots) \supset (\ldots\alpha\ldots)) \supset ((\ldots) \supset (\forall\alpha)(\ldots\alpha\ldots))\text{'}$$

in which '$(\ldots)$' contains no schematic letters and '$(\ldots\alpha\ldots)$' contains only 'α' as a schematic letter. Further, if this proposition is false, then both its antecedents will be true and its consequent will be false. Now since '$(\ldots\alpha\ldots)$' contains only 'α' as a schematic letter, the consequent states that every proposition of the form '$(\ldots\alpha\ldots)$' is true, and since the consequent is by hypothesis false there must be some proposition of the form '$(\ldots\alpha\ldots)$' which is not true. But if this proposition is expressed by, say, '$(-)$', then since '$(-)$' is of the form '$(\ldots\alpha\ldots)$' evidently '$(\ldots) \supset (-)$' is of the form '$(\ldots) \supset (\ldots\alpha\ldots)$', and by the first antecedent every such conditional proposition is true. Hence '$(\ldots) \supset (-)$' is true, but by the second antecedent '$(\ldots)$' is true, and so we must conclude that '$(-)$' is after all true and not false. The hypothesis therefore leads to a contradiction, and so we may conclude that no proposition of the form

$$`(\forall\alpha)(P \supset (-\alpha-)) \supset (P \supset (\forall\alpha)(-\alpha-))\text{'}$$

is false, or in other words that

$$\vdash((\forall\alpha)(P \supset (-\alpha-)) \supset (P \supset (\forall\alpha)(-\alpha-))).$$

Now I have been offering arguments for the fundamental rules and axioms of quantificational logic, and this may suggest that I have been trying to *establish* these rules and axioms, as it were 'out of nothing'. This is obviously a silly aim, for indeed a very cursory inspection will make it clear that many of the principles of argument I have been employing are themselves principles which are usually established as *consequences* of these rules and axioms. But the principles of argument I employ do seem to me very evident ones, and so one thing that I hope is achieved by these arguments is that the fundamental rules and axioms of quantification theory come to be very evident themselves. Further, I have of course been employing in these arguments the account just given of the universal quantifier and of what it is to be a theorem, and so another thing that I hope is achieved by these arguments is that the adequacy and appropriateness of that account is confirmed: if the account is acceptable then these rules and axioms are also acceptable, but it is not clear why the rules and axioms should be acceptable if the account is not. But what I mainly wanted to establish is that these rules and axioms do indeed hold for any variety of quantification whatever. At no point have I made use of any special features which α-components must be supposed to have, and my arguments apply equally whether they are taken to be subject-components, predicate-components, proposition-components, or any other sort of propositional components. So far, then, the claim that quantification theory is the same for all types of propositional components is upheld.

In so far as there are differences, they connect with the further rule I mentioned, and which I have not yet argued for, namely the rule allowing for the intersubstitutibility of different letters from the same alphabet. (Indeed, I have so far been making do with only one α-letter, namely 'α'.) The basic principle underlying this rule can be formulated quite generally, and quite self-evidently, as a part of thesis (*b*) above, namely

(iv) If every proposition of the form '(...)' is also of the form '(—)', then if ⊢(—) then ⊢(...)

and this, as is evident from my discussion in section 1, is just the principle behind the rule of substitution. First, one can easily see the rule allowing for interchange of different letters from the same alphabet as a consequence of this principle, for it results from our conventions for the use of α-letters in representing propositional forms that one α-letter is in general as good as another so long as repetitions of the same letter are preserved. Thus in the case of subject-letters any proposition of the form, say, 'a = b' is also of the forms 'b = c', 'c = a', etc., and conversely, while any proposition of the form 'a = a' is also of the form 'a = b' but not conversely. So we may set up as an auxiliary rule to principle (iv):

(iv)*a* If the formula '(...)' is obtained from the formula '(—)' by replacing all free occurrences of some α-letter in '(—)' with free occurrences of some other α-letter, then every proposition of the form '(...)' is also of the form '(—)'.

This auxiliary rule obviously holds for all types of quantification, and once we have added it to the principles (i)–(iii) already established we do have a basis sufficient for the whole of what is ordinarily called first-order quantification theory. What we do not have, however, is a basis sufficient for quantification over *any* type of propositional component, for what is peculiar to quantification over subject-components is precisely that the auxiliary rule (iv)*a* is the *only* auxiliary rule of this nature that needs to be adopted. If we have two formulae which contain only subject-letters as schematic letters, then the *only* way in which it can happen that every proposition of the first form is also of the second form is that the first form be obtainable from the second by a substitution licensed by (iv)*a*. However, this is evidently not true for other propositional components: it is for instance a fact that every proposition of the form 'P v Q' is also of the form 'P', and that every proposition of the form 'Fa v Ga' is also of the form 'Fa', but these facts are not implied by rule (iv)*a*.

Now I shall not delay here to formulate the auxiliary rule that would be needed for quantifying over propositions, since I do not use such quantification, but I should perhaps say something about quantification over predicates. The important difference here between subjects and predicates is, roughly speaking, that if we take any proposition which contains a subject-component and simply delete that subject-component, then what we have left will be a predicate-component, while on the other hand it is by no means the case that if we take any proposition which contains a predicate-component and delete that predicate-component then we are left with a single subject-component. We may put this point by saying that the notion of a predicate, as it appears in logical theory, is essentially the notion of a *remainder*, and generally we may say that Φ-components are *remainders* to α-components if every proposition which contains an α-component also contains a Φ-component, and may be analysed as containing just those two components put together in the simplest possible form. This simplest possible form for obtaining a proposition from an α-component and a Φ-component will generally be represented by an α-letter and a Φ-letter in direct juxtaposition, as in 'Φα', and it may be called the *elementary* Φ-α form (or sometimes the elementary Φ-form). It is, then, characteristic of the notion of a remainder that if Φ-components are remainders to α-components then every proposition of any form '(—α—)' is also of the elementary form 'Φα', and indeed that it is of these *related* forms (provided 'Φ' does not occur in '(—α—)'). For it is the very same α-component that needs to be supplied to the form '(—α—)' and to the form 'Φα' if we are to obtain our original proposition from each.

To say, then, that Φ-components are remainders to α-components is just to say that every proposition of any form '(—α—)' is also of the related form 'Φα',[8] and from this it follows that any proposition whose form is some *complication* of '(—α—)' is also of the form which is the *corresponding complication* of 'Φα'. This is a rough statement of the auxiliary rule that we require here, but to formulate it more exactly we shall need to be more precise over the notion of a complication and a corresponding complication. To this end, let us say that in any whole formula which contains an α-letter occurring freely we can recognize a Φ-*schema*, which is what is obtained by deleting some free occurrences of that α-letter so as to leave gaps, and which may be *completed* with the α-letter just deleted so as to yield the original formula. And let us call a Φ-schema which consists of a single Φ-letter and a gap an *elementary* Φ-schema. Now any formula which contains several occurrences of a Φ-letter is a *complication* of the elementary Φ-schema answering to that Φ-letter, though we may note that the Φ-schema may be completed by different α-letters at its different occurrences, and these various α-letters may occur bound or free. And if we now take all free occurrences of that Φ-letter, and replace the elementary Φ-schema it represents by a new Φ-schema, which is completed at each of its occurrences with the same α-letter as was the corresponding

[8] Provided 'Φ' does not occur in '(—α—)'. I shall generally leave this proviso unstated in future.

occurrence of the original elementary Φ-schema, then we shall have formed the *corresponding* complication of the new Φ-schema.

To this explanation we should add the provisos that whatever letters occurred freely in the newly introduced Φ-schema considered on its own must continue to occur freely in the formula as a whole at all occurrences of the newly introduced Φ-schema, and further that the α-letters which occurred free or bound to a certain quantifier where they completed the original elementary Φ-schema must continue to occur free or bound to the same quantifier where they complete the newly introduced Φ-schema. The first proviso here could be omitted if we were retaining a notational distinction between schematic letters and variables, for the schematic letters in a Φ-schema would then be incapable of being captured by quantifiers in the formula into which it was introduced, since our notational conventions would prevent quantifiers from binding schematic letters at all. And both provisos could be omitted if we were using not the usual quantifier notation but the arrow-notation suggested on p. 58, for the point of the proviso is just to prevent the bondage-links of the two formulae being disturbed when the one is inserted into the other, and if the bondage-links are explicitly drawn in beforehand then such a disturbance cannot occur. It is only because repetition of letters is a less explicit way of indicating bondage-links that it can lead us astray in this instance. We may say, then, that the provisos are needed only because we have sacrificed some relevant notational distinctions for the sake of a simpler and more economical presentation of the whole, and this shows, I think, that their presence is of no philosophical interest.

Anyway the auxiliary rule that we need for the theory of quantification over Φ-components can now be stated as

(iv)*b* If the formula '(...)' is obtained from the formula '(—)' by replacing all free occurrences of some elementary Φ-schema in '(—)' by freedom-preserving occurrences of some other Φ-schema, then every proposition of the form '(...)' is also of the form '(—)'.

This rule applies wherever Φ-components are remainders to α-components in the way explained,[9] and its justification is

[9] Note that although no reference to α-components occurs explicitly in rule (iv)*b*, it does occur in the definition of 'Φ-schema'.

easily enough seen with the help of thesis (*a*) on p.65. Evidently we could go on to generalize the rule so that it could be used for quantifying over polyadic predicates generally, and not just monadic predicates, but in fact the construction of arithmetic which I sketched in the appendix to chapter 1 made no use of such quantification. Indeed, apart from the manœuvres designed to yield Frege's theorem (which I shall discuss later), the only applications of rule (iv)*b* in that construction were steps where an expression such as '$\mathrm{Fx}\ \&\ \mathrm{x}\neq\mathrm{a}$' was substituted for 'Fx', and it is easy enough to see that the rule does indeed license these steps.

Consider for example the proof of theorem 2 in that appendix, where I first set up the hypothesis

$$(\forall \mathfrak{f})((\exists \mathrm{nx})(\mathfrak{f}\mathrm{x}) \supset (\exists \mathrm{mx})(\mathfrak{f}\mathrm{x}))$$

and at once deduce from this

$$(\exists \mathrm{nx})(\mathrm{Fx}\ \&\ \mathrm{x}\neq\mathrm{a}) \supset (\exists \mathrm{mx})(\mathrm{Fx}\ \&\ \mathrm{x}\neq\mathrm{a}).$$

In detail the reasoning is that we first set down the theorem

$$(1): \quad (\forall \mathfrak{f})((\exists \mathrm{nx})(\mathfrak{f}\mathrm{x}) \supset (\exists \mathrm{mx})(\mathfrak{f}\mathrm{x})) \supset ((\exists \mathrm{nx})(\mathrm{Fx}) \supset (\exists \mathrm{mx})(\mathrm{Fx}))$$

which is justified by the rule of instantiation (rule (ii)) applying to all varieties of quantification theory. Then we observe that the formula

$$(2): \quad (\forall \mathfrak{f})((\exists \mathrm{nx})(\mathfrak{f}\mathrm{x}) \supset (\exists \mathrm{mx})(\mathfrak{f}\mathrm{x})) \supset ((\exists \mathrm{nx})(\mathrm{Fx}\ \&\ \mathrm{x}\neq\mathrm{a}) \supset (\exists \mathrm{mx})(\mathrm{Fx}\ \&\ \mathrm{x}\neq\mathrm{a}))$$

is easily obtained from this formula by predicate-substitution, or in other words that (2) is the same complication of the predicate-schema '$\mathrm{F}...\ \&\ ...\neq\mathrm{a}$' as (1) is of 'F...', and so by rule (iv)*b* we conclude that every proposition of the form (2) is also of the form (1). Consequently we infer that since (1) is a theorem then by rule (iv) (2) must be a theorem, and the fact that (2) is a theorem is what licenses the deduction in question.

I have been through this in detail because I want to make it quite clear that although in this argument we have gone beyond the principles needed for first-order quantification theory, nevertheless the argument must be regarded as being 'purely logical' in nature, and further it must be admitted to be completely free from any assumptions about the existence of *objects*, such as classes. In asserting that every proposition of the form (2) is also of the form (1) we do *not* rely on the principle that if there is such a class as the class of all F-things then there is equally such a class as the class of all F-things other than a. The existence or non-existence of classes has absolutely nothing to do with rule (iv)*b*, or with any other part of the theory of quantification over predicates.

3. Quantification and Existence

It has of course been my main object in presenting this account of quantification to make it clear that one may introduce quantifiers ranging (as we say) over predicates without at all treating predicates as objects, and similarly for any other variety of quantification. By way of completing the account I shall now contrast it explicitly with Quine's position on quantification, and consider what bearing it has on his dictum that to be is to be the value of a bound variable.

It seems to me that there are three main considerations which lead Quine to espouse this dictum: one is that quantification into 'intensional' or 'referentially opaque' contexts cannot be construed, and therefore there is no call to countenance non-existent 'intensional' entities as values of our variables;[10] another is that it is altogether more convenient to banish names or logical subject-expressions from our 'canonical notation', and therefore there is no call to use anything other than the existential quantifier to make existence-claims;[11] and the third is that the existential quantifier can *only* be construed as a device for making existence-claims, for attempts to read it differently are all somehow illegitimate. Evidently it is this last contention that comes closest to conflicting with my position, and what is certainly true is that the argument Quine gives for this last contention is directly opposed to what I have been urging; for what he argues for is primarily that to quantify over things of any sort must be to treat those things as *objects*, and it is on this point that he bases his further conclusions about *existence*.

Consider, for instance, the opening stages of his argument in *Methods of Logic:*

The evident analogy between variables 'x', 'y', etc., and schematic

[10] See especially his 'Notes on Existence and Necessity', *Journal of Philosophy*, 1943.

[11] *From a Logical Point of View*, chapter 1; *Word and Object*, section 37.

letters 'F', 'G', etc., tempts us to try using the latter in quantifiers, e.g., thus:

(1) $(F)[(x)Fx \supset (\exists x)Fx]$,

(2) $(\exists F)[(\exists x)Fx . -(x)Fx]$.

However, let us not be hasty in supposing that we understand (1) and (2). We have been reading the quantifiers '(x)' and '(∃x)' in the fashion 'each thing x is such that' and 'something x is such that', but how are we to read '(F)' and '(∃F)'? May we read '(F)' in the fashion 'each general term (or predicate) F is such that', and '(∃F)' correspondingly? No, this is a confusion. 'F' has never been thought of as referring to general terms (and thus as standing in place of *names of* general terms), but only as standing in place of general terms. If there were objects of a special sort, say gimmicks, of which general terms were names, then the proper readings of '(F)' and '(∃F)' would be 'each gimmick F is such that' and 'some gimmick F is such that'. But the difficulty is that general terms are not names at all. (*Methods of Logic*, § 38, p. 225)

Clearly then Quine claims that in introducing quantifiers ranging over predicates one must be implicitly treating predicate-expressions as names (and therefore one must be treating predicates as objects), and so far as one can see his ground for this claim is simply the assertion that quantified formulae of the sort

$$(\forall\alpha)(-\alpha-)$$
$$(\exists\alpha)(-\alpha-)$$

must be rendered into English in the fashion

Each ... is such that — it —

Some ... is such that — it —

when the first blank is filled by a noun of classification such as 'thing', 'general term', or 'predicate', appropriate to the type of variable in question. Now I would say that if Quine is granted this premiss then his position is virtually unassailable, but I do not see why the premiss should be granted. Quine offers no arguments in its support, so far as I am aware, though I have offered an argument in favour of my alternative proposal, namely that it is adequate for, and I think required by, perfectly ordinary quantification theory. In this theory a subject–predicate formula such as 'Fa' is a legitimate substitution for an elementary propositional formula such as 'P', and thus from the fact that

'P ∨ ∼P' is a theorem it follows that 'Fa ∨ ∼Fa' is a theorem. With my account of what it is to be a theorem[12] the justification for this mode of inference is clear, but if the assertion that 'Fa ∨ ∼Fa' is a theorem means anything importantly different from what I suggest, then its justification is still to seek. Again in this theory a subject–predicate formula is a theorem if and only if its universal closure is a theorem, and thus from the fact that 'Fa ∨ ∼Fa' is a theorem it follows that '(∀x)(Fx ∨ ∼Fx)' is a theorem, and conversely. I find it intensely difficult to see why this should be so if it is not fitting to explain the universal quantifier in terms of what it is to be a theorem in the way I have suggested. At the very least, my reading of the quantifiers must surely be *a* permissible reading for the quantifiers that occur in ordinary first-order quantification theory, namely quantifiers binding subject-variables, and there is therefore no reason to reject it in the case of other quantifiers.

With this reading of the quantifiers one might well say that all occurrences of quantification as ordinarily understood are reduced to quantification over propositions,[13] and one might perhaps add that to come to a proper understanding of this we shall need to know *what propositions there are*. This is no doubt a fair comment, but one which would require for its discussion a full account of what a proposition is, and this I do not intend to give. On this point I said (in chapter 2, section 3) that all that I required of an account of propositions was that it should yield the appropriate connection between propositions and truth and falsehood. In particular I required that every proposition be either true or false and that no proposition be both true and false, since this is the essential basis of the propositional calculus. But I also required that it should always be possible to mention propositions, and it has now emerged that this latter requirement is a requirement of some magnitude, when understood as I wish it to be understood. For the requirement may be put by saying that wherever we should want to speak of a truth or a falsehood

[12] With, if necessary, the further modal complication noticed in the text to note 3.

[13] 'Thus generally: "$(x)(\phi x)$" is to mean "ϕx always". This may be interpreted . . . as "all propositions of the form ϕx are true" or "all values of the function ϕx are true". Thus the fundamental *all* is "all values of a propositional function", and every other *all* is derivative from this.' (Russell, 'Mathematical Logic as based on the Theory of Types', § 3, reprinted in *Logic and Knowledge*, p. 72.)

there should *be* some proposition which we can say is what is true or false in this case. This has the consequence that an account of propositions will be intuitively unacceptable if it implies that what propositions there are depends upon what sentences some language or other happens to contain, or on what speech-acts there happen to have been, or on what anybody has actually thought, for we do not commonly suppose that we create *new truths* by adding to our vocabulary or by speaking and thinking on novel topics. Propositions may indeed be explicable in terms of sentences, speech-acts, or thoughts—that is a matter which I do not intend to argue here—but if so it must be in some appropriate sense *possible* sentences, speech-acts, or thoughts that are ultimately invoked. Of course theoremhood and quantification will be perfectly well defined for any account of propositions which satisfies my first requirement, but the definition will not define what it is meant to unless the second requirement is satisfied too.

I do not think it can reasonably be objected to me that in taking quantification over propositions as the fundamental type of quantification (for logical purposes) I have merely obscured what was previously perfectly clear, for I would not agree that the question of what propositions there are is any more obscure than the Quinean question of what *things*, *entities*, or *objects* there are. (It should be observed that, as Quine understands this question, it includes the question of what *classes* there are, and in view of Russell's paradox that question is certainly not one that we already know how to answer—as Quine fully realizes.) Besides, however obscure the notion of 'every proposition' may be, my argument has been that it is needed even in the earliest stages of logic if we are to explain the significance of any formula's being a theorem (or axiom). It may legitimately be said that when quantification is introduced we progress from stating that every proposition is thus and so to considering statements that *not* every proposition is thus and so,[14] but I just do not see how it can be maintained that the notion of 'every proposition' is clear while the notion of 'not every proposition' remains obscure. It may also be said that when quantification is introduced we progress from speaking of every proposition of a certain form to

[14] This is intended as a concise formulation of Russell's distinction between 'all' and 'any'.

speaking of every proposition of a certain form *with respect to* some schematic letter, and it could reasonably be complained that I have not yet given a proper account of this locution. I have given examples and given metaphors, but that might appear to leave the ontological implications of the locution somewhat unclear.

Let me recall, then, that what needs explanation is the use of formulae containing quantifier-expressions, i.e. the use of formulae containing symbols of the sort '(∀α)(—α—)'. For definiteness, I shall now go through this explanation in connection with quantification over subjects, since that is a matter which I left unexplained at the end of the last chapter, but it will be clear how to transfer the account to any other variety of quantification. What needs to be explained, then, is what it means to say that a proposition is of the form '(∀x)(Fx)' or '(∀x)(Fa ∨ Gx') or '(∀y)(∀x)(Fy ∨ Gx)', and so on; and what it means to say that several propositions are of related forms of this sort. Let us take the second case first. Considering then the forms 'Fa' and '(∀x)(Fx)' we may say that two propositions are of these related forms if and only if the first *contains some predicate initially* and the second is the proposition that every proposition is true which *contains the same predicate initially* as the first does. Since two propositions contain the same predicate initially if and only if they are of the *related* forms 'Fa' and 'Fb', another way of putting this is to say that the first proposition must be of the form 'Fa' and the second proposition must be the proposition that all propositions of the *related* form 'Fb' are true, and this way of putting it has the advantage that it is easily extended to more complicated cases. For where we had previously spoken of a form *with respect to* some schematic letter,we can now achieve just the same effect by exhibiting two formulae which differ just in that schematic letter. So for instance we may say that two propositions are of the related forms 'Fa ∨ Gb' and '(∀x)(Fa ∨ Gx)' if and only if the first is of the form 'Fa ∨ Gb' and the second is the proposition that all propositions of the *related* form 'Fa ∨ Gc' are true, and similarly for other cases. In this way we may explain any pair of related forms where the one is obtained by prefixing a quantifier to the other, and we may now use this to explain what it means to say of some one proposition that it is of the form '(∀x)(Fx)' or generally of any form

'$(\forall x)(-x-)$'; for I explain this by saying that it is for that proposition and some other proposition to be of the related forms '$(\forall x)(Fx)$' and 'Fa', or generally '$(\forall x)(-x-)$' and '$(-a-)$' (—where 'a' does not occur in '$(\forall x)(-x-)$'). And by employing both these principles together we may further explain any other sets of related forms. For instance we may say that two propositions are of the related forms '$(\forall x)(Fa \vee Gx)$' and '$(\forall x)(Fx \vee Hb)$' if and only if there are two further propositions of the related forms '$Fa \vee Gc_1$' and '$Fd_1 \vee Hb$', and our first proposition is the proposition that all propositions of the further related form '$Fa \vee Gc_2$' are true, while our second proposition is the proposition that all propositions of the yet further related form '$Fd_2 \vee Hb$' are true. To understand this explanation all that we need to be told beforehand is when four propositions are of the related forms '$Fa \vee Gc_2$' '$Fa \vee Gc_1$', '$Fd_1 \vee Hb$', '$Fd_2 \vee Hb$', and this has indeed already been catered for in chapter 2. Further, chapter 2 explained truthfunctional combination of formulae quite generally, and so we shall now also be in a position to explain, say, the related forms '$\sim(\forall x)(Fa \vee Gx)$' and '$P \supset (\forall x)(Fx \vee Hb)$', and from this position we can further explain the related forms '$(\forall y)(\sim(\forall x)(Fy \vee Gx))$' and '$(\forall y)(P \supset (\forall x)(Fx \vee Hy))$', and so on. Evidently all the uses of related forms that we shall require can be explained in this fashion.

This will complete the explanation of formulae of the monadic predicate calculus. The basic notions that need to be explained are the notions of containing a subject, containing the same subject, and containing the same predicate initially; or in other words the basic points to be explained are what it means for a proposition to be of an elementary subject–predicate form, such as 'Fa', what it means for several propositions to be of such related forms as 'Fa', 'Ga', 'Ha',..., and what it means for several propositions to be of such related forms as 'Fa', 'Fb', 'Fc',.... Given an understanding of the negation of a proposition, and of the conjunction, disjunction, etc. of two propositions, we can then explain all formulae built up by truth-functional combination from any formulae we have explained already, and given an understanding of 'Every proposition' we can further explain the universal quantification of any formula we have explained already. All this yields an explanation of all the formulae of the monadic predicate calculus, and it yields more as well; for the explanation of quantification over subjects applies unchanged to quantification over monadic predicates.

To return, finally, to the topic of quantification and existence, I have been arguing that we are quite able to introduce quantification over so-and-so's without treating so-and-so's as *objects*, and therefore we are in a position to reject an important part of Quine's argument for the dictum that to be is to be the value of a bound variable. I have not, however, argued that there is anything wrong with the dictum itself, and I would be perfectly sympathetic to a revised version of it to the effect that to be is to be is to be the value of a bound *subject*-variable. Since all Quine's variables *are* subject-variables this is at least partially in accord with his intentions, and it can be supported on the basis of my analysis if it can be argued that for a subject-predicate proposition to be *true* its subject must *exist*. In the case of other propositional components there seems to be an insuperable difficulty with the dictum in that the sentence-frame '. . . exists' must apparently be completed by an expression which introduces a propositional *subject*, but if this difficulty could somehow be overlooked I would have no objection to the view that, e.g.,

Some proposition of the form 'F(Socrates)' is true

may be taken as asserting the existence of something that is true of Socrates. (Though I would want to suggest that in that case 'something' should not be taken to be equivalent to 'some thing', but assimilated rather to the indissoluble expressions 'somehow' and 'somewhere'.) But my main point is that even if we do suppose that such a proposition asserts the existence of something true of Socrates, we must not take it as asserting the existence of any *object* that is true of Socrates. Whether we can make sense of the existence of something not an object is a question that may be left on one side, at least for purposes of logic, for as I have said before what is important for logical purposes is not what can be said of the propositional components themselves, but what can be said of propositions containing those components.

In this connection I should draw attention to one feature of my explanation of formulae involving quantifiers, and that is that there can be no propositions of any quantified form unless there are propositions of the form therein quantified. For instance I explained that a proposition is of the form '$(\forall x)(Fx)$' if and only if it and some other proposition are of the related forms '$(\forall x)(Fx)$' and 'Fa', and therefore there will be no propositions of the form '$(\forall x)(Fx)$' unless there are propositions of the form 'Fa'. This is quite deliberate, for it can be seen independently that in electing to quantify over a certain type of propositional component, and in making use of ordinary quantificational logic in connection with this type of quantification, we do anyway commit ourselves to the existence of propositions containing these components. It is commonly remarked that orthodox quantification theory is committed to the assumption that at least something exists insofar as it contains theorems such as '$(\forall x)(Fx) \supset (\exists x)(Fx)$', and in my version this assumption is the assumption that there are at least some subject-predicate propositions. Now a scrutiny of the arguments I gave in section 2 for the rules and axioms of quantification theory will not reveal any such assumption, for in my version the assumption enters rather when the rules of inference of propositional logic are taken to apply unchanged to formulae of quantification theory. This may be seen by considering, for example, the rule Modus Ponens, which we may formulate as

> if every proposition of the form '(—)' is true, and every proposition of the form '(—) ⊃ (...)' is true, then every proposition of the form '(...)' is true also.

If we attempt to frame a justification for this rule to parallel the justifications I gave in section 2 we shall find that it is necessary to assume that there are some propositions of the form '(—)'; for if there were no propositions of this form, and consequently no propositions of the form '(—) ⊃ (...)', then both the antecedents of this rule would be true, though the consequent could well be false if there were propositions of the form '(...)'. And in fact it will be found that any attempt to deduce a theorem such as '$(\forall x)(Fx) \supset (\exists x)(Fx)$' from the rules and axioms I have stated would involve an application of Modus Ponens (or some similar rule) which would require for its validity the assumption that there are propositions of subject–predicate form.

But it should be observed that this existential commitment is still not, in fact, a commitment that arises particularly through the use of *quantifiers*, but is actually a consequence of taking the rules of propositional logic to apply to formulae containing *schematic letters* for the components in question. For suppose that there were no propositions of the form 'Fa'. In that case there would be no propositions of the forms 'P ⊃ Fa' or 'Fa ⊃ Q' either, and so we could truly say that all propositions of the form 'P ⊃ Fa' are true, and similarly that all propositions of the form 'Fa ⊃ Q' are true. The usual rules of propositional logic will then allow us to deduce—e.g. by using Modus Ponens twice on the tautology '⊢(P ⊃ Fa) ⊃ ((Fa ⊃ Q) ⊃ (P ⊃ Q))'—that all propositions of the form 'P ⊃ Q' are true also (and hence that every proposition is true). But clearly the inference must be illegitimate if the two premisses are asserted as true only because they are vacuously true. The reason why the difficulty appears to enter with the advent of quantified formulae, and not merely with the use of schematic letters, is that it is only with the advent of quantified formulae that we are actually given as true any formulae of the pattern '(—) ⊃ Fa' and 'Fa ⊃ (---)' which lack some of the schematic letters of 'Fa' in the gaps displayed, namely the formulae '(∀x)(Fx) ⊃ Fa' and 'Fa ⊃ (∃x)(Fx)'. So this existential assumption is indeed required by orthodox quantification theory, but only because it is really required at an earlier stage by the introduction of schematic letters for the relevant components; and anyway the assumption is surely an innocuous one.[15]

[15] The next chapter will draw attention to a further assumption that is required if the usual rules of propositional logic are taken to apply without restriction to formulae with new schematic letters.

4. Quantifiers and Types

1. Subject-Quantifiers

In the last chapter I argued that to quantify over any type of propositional component is not to commit oneself to any further assumptions than are involved in the use of schematic letters for those components. Once the schematic letters are properly introduced the reasoning which justifies the application of quantification theory is essentially the same no matter what type of propositional component is in question, and it is reasoning which must surely be admitted to be of a purely logical nature. To apply this to the case of quantification over numbers, then, all that is strictly needed is that we should explain the use of schematic letters to stand in place of numerals, which will depend upon our explaining their use in expressions for numerical quantifiers, and then this part of the construction at least will be fully justified. Now it would be possible to explain the notion of a numerical quantifier directly, but I think that the situation becomes clearer if numerical quantifiers are introduced as a special case of something yet more general, which I call the general notion of a quantifier. Besides, as I mentioned in chapter 1, this method of approach should enable us to introduce numerical quantifiers by *definition*, and the definition should provide the relevant premiss for a proof of the two theses adopted in chapter 1 as axioms. The object of this chapter is, then, to introduce the general notion of a quantifier, but I begin in this section with the more restricted notion of a quantifier which binds *subject-variables*, or as we may call it a *subject*-quantifier.

What I have in mind here is pretty well what Frege meant by a second-level concept, and what others have meant by a second-level predicate, for the main point is that subject-quantifiers are *remainders* to (first-level) predicates in just the way in which (first-level) predicates are *remainders* to subjects (cf. pp. 69–70). Speaking very roughly, we may say that a subject-quantifier is what remains of any proposition containing a predicate when some occurrences of that predicate are dropped out, just as a predicate is what remains of any proposition containing a subject when some occurrences of that subject are dropped out, and accordingly I shall form an expression for a quantifier by taking an expression for a proposition (a sentence) and leaving a gap in place of its predicate-expression just as we standardly form an expression for a predicate by taking an expression for a proposition and leaving a gap in place of its subject-expression. Thus for instance the expression '(∀x)(—x—)' will be an expression for the universal quantifier (as applied to subject-variables).

I should hasten to point out that this notion of a quantifier is very much more general than one might intuitively have expected. Later in this chapter I shall introduce a more restricted conception of a quantifier, but at present not only are '(∀x)(—x—)' and '(∃x)(—x—)' expressions for quantifiers, and of course '(4x)(—x—)' and '(∃4x)(—x—)', but so also are such apparently hybrid expressions as

'(∀x) (x is a man ⊃ —x—)'

'Socrates is wise ⊃ (∃x)(—x—)'

'(∀x)(—x—) ⊃ (∃x)(—x—)'

and even '(—Socrates—)'.

One might be inclined to object to the second example just displayed that it is only half of the expression that is a quantifier-expression, and to the third that it contains two quantifier-expressions but is not as a whole one, but I think that such objections would be unpersuasive. For we do not find any difficulty in recognizing e.g. the expressions

'Socrates is a man ⊃ ... is mortal'
'... is a man ⊃ ... is mortal'

as predicate-expressions, and yet their relation to a paradigm predicate-expression is quite analogous to the relation of our two expressions to a paradigm quantifier-expression. Besides, it would lead to most unwelcome results if we tried to limit quantifier-expressions to those in which some recognized quantifier occurred at the beginning with the whole of the rest of the expression in its scope, for then it would seem that even the existential quantifier would no longer count as a quantifier if it were defined by the expression '$\sim(\forall x)(\sim(-x-))$'. With the last example, however, it would seem that I can no longer say even that what is here recognized as a quantifier-expression is simply a logical complication of what would ordinarily be recognized as a quantifier-expression, for '(—Socrates—)' seems merely to be a strange subject-expression and surely subjects are not quantifiers at all.

However, it seems to me that even this can be made acceptable at an intuitive level. Though I shall draw an important distinction later, we may for the moment say that a predicate can always be regarded as a propositional form which calls for a subject to complete it into a proposition, and the leading idea in my notion of a quantifier is that a quantifier can always be regarded as stating of a propositional form that certain ways of completing that form into a proposition yield a true proposition. Thus the universal quantifier and the existential quantifier can certainly be regarded as stating that *all* or that *some* ways of completing the form yield a true proposition, and it is easy enough to see a numerical quantifier as stating *how many* ways of completing the form yield a true proposition. But similarly the quantifier I intend by '(—Socrates—)' can be seen as stating that one particular way of completing the form yields a true proposition, namely completing it with the subject here expressed by 'Socrates'. Accordingly I think it would be reasonable to regard the expression '(—Socrates—)' as a short way of writing

It is true of Socrates that —he—,

in just the way that it is reasonable (in the case of subject-quantification) to regard '$(\forall x)(-x-)$' as a short way of writing

It is true of every object that —it—,

provided that we make no distinction between the propositions expressed by the two sentences

Socrates is wise
It is true of Socrates that he is wise.

One might say that the second sentence here formulates the proposition in a way that helps us to see that it contains the quantifier '(— Socrates —)', just as at a different level one might perhaps say that of the two sentences

> If Socrates is a man then Socrates is mortal
> Socrates is, if a man, then mortal

the first formulates the proposition in a way that helps us to see that it is of the form 'P ⊃ Q', while the second formulates it in a way that helps us to see that it contains the *one*-place predicate '... is a man ⊃ ... is mortal'.

Now the question arises whether the subject expressed by 'Socrates' and the associated subject-quantifier expressed by '(— Socrates —)' need to be distinguished in any way, for this point is not decided by the fact that any proposition containing the first contains the second also. The only evident difference between them is that the quantifier has a *scope* (expressed by its outer brackets) while the subject does not, and perhaps it will be easier to discuss this question if we explicitly rewrite the quantifier as

(Socrates = x)(— x —)

(to be read as 'For Socrates as x, — x —') so as to represent this scope more vividly.[1] Now if there is ever any reason to suppose that a subject-expression has an effective scope, that will furnish ground for saying that the proposition in question may not in fact contain a subject at all, but only its associated subject-quantifier. So far as I can see, the only plausible cases here are those involving intensionality. For instance, to adapt one of Geach's examples,[2] it would certainly appear that we should distinguish between

(∃x)(Johnson believes that x is a shopkeeper)

and Johnson believes that (∃x)(x is a shopkeeper),

but similarly it might well be urged that we should distinguish between

(Ralph de Vere = x)(Johnson believes that x is a shopkeeper)

and

Johnson believes that (Ralph de Vere = x)(x is a shopkeeper),

[1] One of the ways in which Russell's definite descriptions differ from subjects proper is in having scopes, and a similar notation can be used with advantage in their case.

[2] See *Reference and Generality*, § 94.

for perhaps we might count the first true and the second false in the case where Johnson does believe of some man that he is a shopkeeper, and that man is in fact Ralph de Vere, but Johnson does not know or even suspect that he is. Well, for my present purposes I do not think it necessary to settle this very awkward question, but only to point out that I am not settling it. Any proposition which contains a subject will also contain its associated subject-quantifier, and it seems safe to rule that any proposition which contains that subject-quantifier *initially* (i.e. in such a way that its scope is that whole proposition) may equally well be seen as containing the relevant subject. Thus we shall be able to assert

$$\vdash \mathrm{Fa} \equiv (\mathrm{a}{=}\mathrm{x})(\mathrm{Fx})$$

and from this it will apparently[3] follow that

$$\vdash \text{b believes that Fa} \equiv (\mathrm{a}{=}\mathrm{x})(\text{b believes that Fx}).$$

But owing to the intensionality of the belief-construction we need *not* accept as a further consequence

$$\vdash \text{b believes that Fa} \equiv \text{b believes that } (\mathrm{a}{=}\mathrm{x})(\mathrm{Fx}),$$

and this will allow us to preserve the appropriate distinction if we think it is worth preserving.

I have so far been addressing myself only to the question whether we should maintain that a proposition contains a subject if and only if it contains the associated subject-quantifier. If this question is answered affirmatively, it might then seem that we should be faced with the further question whether the subject which a proposition contains is itself precisely the same thing as the associated subject-quantifier that it contains, but in fact I see no need to consider this further question at all. Although I have been speaking so far, in an informal way, as if there were certain entities called subject-quantifiers (and named by quantifier-expressions) there is no more need to make this assumption here than there was in the case of subjects and predicates. For we shall not need to consider quantifiers in isolation from their occurrence in propositions, and therefore we shall not need to say what a subject-quantifier *is*; it will be quite sufficient if we explain what it means to say that a proposition

[3] I say 'apparently' because of course even this inference can be denied, if we refuse to allow such expressions as 'Johnson believes that . . . is a shopkeeper' as genuine predicate-expressions. For a further aside on this aspect of intensionality see chapter 5, note 2.

contains a subject-quantifier, and what it means to say that several propositions *contain the same subject-quantifier*, for—roughly speaking—it is only this that is required if we are to make use of the relevant formal notation.

More precisely, then, I first introduce the series of signs '$\mathfrak{Q}$', '$\mathfrak{Q}_1$', '$\mathfrak{Q}_2$',..., as the series of schematic quantifier-letters, associating with it the series '$\mathfrak{q}$', '$\mathfrak{q}_1$', '$\mathfrak{q}_2$',..., of quantifier-variables. Retaining 'x', 'y', 'z',..., as subject-variables I then form the elementary subject-quantifier schemata

$$\begin{array}{ll} (\mathfrak{Q}\mathrm{x})(\text{—}\,\mathrm{x}\,\text{—}), & (\mathfrak{Q}\mathrm{y})(\text{—}\,\mathrm{y}\,\text{—}), \ldots \\ (\mathfrak{Q}_1\mathrm{x})(\text{—}\,\mathrm{x}\,\text{—}), & (\mathfrak{Q}_1\mathrm{y})(\text{—}\,\mathrm{y}\,\text{—}), \ldots \\ \quad\vdots & \quad\vdots \\ \quad\cdot & \quad\cdot \end{array}$$

which I say may be completed with single predicate-letters 'F', 'G',..., to form elementary subject-quantifier formulae of the sort '($\mathfrak{Q}$x)(Fx)', ($\mathfrak{Q}$y)(Gy)', '($\mathfrak{Q}_1$x)(Fx)', and so on. An elementary subject-quantifier formula is then a formula representing a propositional form, and I say that a proposition is of such a form if and only if it *contains a predicate somewhere*. Similarly, several propositions are of such related forms as '($\mathfrak{Q}_1$x)(Fx)', '($\mathfrak{Q}_2$x)(Fx)', '($\mathfrak{Q}_3$x)(Fx)',... if and only if they all *contain the same predicate somewhere*; and several propositions are of such related forms as '($\mathfrak{Q}$x)(Fx)', '($\mathfrak{Q}$x)(Gx)', '($\mathfrak{Q}$x)(Hx)',..., if and only if they all contain the same subject-quantifier initially, which in turn will be the case if and only if each of them *contains a predicate wherever and only where* all the others contain a predicate.

Now if we could specify which propositions were propositions that contained a predicate somewhere, there would be no further difficulty about the remaining notions. For we should then have some nicely circumscribed totality of formulae '(—F—)' written with a predicate-letter somewhere of such a sort that we could say that a proposition contains a predicate somewhere if and only if it may be analysed as one of these forms '(—F—)'. In that case we could go on to say that two propositions contain the same predicate somewhere if and only if they may be analysed as of some related forms '(— F—)' and '(... F ...)' written with the same predicate-letter somewhere, and that one proposition contains a predicate wherever and only where another does if and only if they may be analysed as of some related forms '(—F—)' and '(—G—)', where the formula '(—F—)' is obtainable by substituting the letter 'F' for every occurrence of 'G' in '(—G—)', and the formula '(—G—)' is obtainable by substituting the letter 'G' for every occurrence of 'F' in '(—F—)'—in each case leaving the rest of the formula unchanged.[4] Further, if these explanations are to hand there is evidently no further difficulty in explaining such related forms as '($\mathfrak{Q}$x)(Fx)', '(∀x)(Fx)', and 'Fa', or in expanding the account to cover cases in which the predicate-letter 'F' is supplanted by a more complicated predicate-*schema*.[5] We may then use the methods of the previous chapters to explain all further formulae with schematic quantifier-letters governing subject-variables, though perhaps I should add a word about the case where one such quantifier-letter occurs within the scope of another, as for instance in '($\mathfrak{Q}$x)($\mathfrak{Q}$y)(Fx v Gy)'. Here we say that a proposition is of this form if and only if it and some further proposition are of the related forms '($\mathfrak{Q}$x)($\mathfrak{Q}$y)(Fx v Gy)' and '($\mathfrak{Q}$y)(Fa v Gy)', and the idea is that this will be so if and only if the first contains the predicate here represented by '($\mathfrak{Q}$y)(F... v Gy)' wherever and only where the second contains the predicate here represented by '(Fa v G...)'. So, more formally, what is required is that there be some formula '(—(Fa v G...)—)' containing the predicate-schema '(Fa v G...)' somewhere, and another formula

'(—($\mathfrak{Q}$y)(F... v Gy)—)'

containing the predicate-schema '($\mathfrak{Q}$y)(F... v Gy)' in the same way—i.e. in such a way that each formula is obtainable from the other by substituting the one predicate-schema for all occurrences of the other—and then we say that two propositions are of the related forms '($\mathfrak{Q}$y)(Fa v Gy)' and '($\mathfrak{Q}$x)($\mathfrak{Q}$y)(Fx v Gy)' if the first is both of the form '(—(Fa v G...)—)' and of the related

[4] So for instance any pairs of propositions of any of the related forms

'Fa' and 'Ga'
'(∀x)(Fx)' and '(∀x)(Gx)'
'P ⊃ ∼(∃x)(Fx)' and 'P ⊃ ∼(∃x)(Gx)'
'(∀x)(∃y)(xRy & Fy)' and '(∀x)(∃y)(xRy & Gy)'

will also be pairs of propositions of the related forms '($\mathfrak{Q}$x)(Fx)' and '($\mathfrak{Q}$x)(Gx)'.

[5] For an account of predicate-schemata see p. 70.

form '(𝔔y)(Fa v Gy)', while the second is of the further related form '(—(𝔔y)(F... v Gy)—)'. In this way the problem is reduced to one in which no quantifier-letter occurs within the scope of another, and the criterion first given is easily applied.[6]

All these explanations, however, rest upon the initial assumption that we have already understood the fundamental idea of containing a predicate somewhere, and the difficulty in introducing the concept of a subject-quantifier is just the difficulty of reaching an adequate analysis of this notion. Now in chapter 2 I treated of containing a subject (somewhere), and in terms of this explained the notion of containing a predicate *initially*. However, I then went on to give an explanation of all formulae which could be built up from the elementary subject–predicate formulae by truthfunctional combination, and for greater completeness we may also add here truthfunctional combination with elementary propositional formulae and with elementary relational formulae of any degree. In chapter 3 I proceeded with an account of universal quantification which allows us to include all formulae built up from these by applying universal quantification at any point, re-applying truthfunctional combination, re-applying universal quantification, and so on. This provides a great variety of formulae containing schematic predicate-letters in many different positions—e.g. in positions such as '(...) ⊃ (∀x)(Fx v ...)'—and clearly we shall want to say that any proposition which is of one of these forms contains a predicate somewhere. But a little reflection shows that it is not *only* propositions of these forms that contain a predicate somewhere. For once we have succeeded in introducing our elementary subject-quantifier formulae we shall want to be able to subject them to quantification in their turn, so as to obtain formulae such as '(∃𝔮)(𝔮x)(Fx)';[7] and it is clear that any proposition of this form

[6] In effect our proposition is required to be of some form

'(—(—(Fx v Gy)—)—)',

where '(—(Fx v G...)—)' is the same complication of '(Fx v G...)' as '(—(—(F... v Gy)—)—)' is of '(—(F... v Gy)—)'. So for example any proposition of the form 'P ⊃ ∼(∀x)(P ⊃ ∼(∀y)(Fx v Gy))' is also a proposition of the form '(𝔔x)(𝔔y)(Fx v Gy)', and so is any proposition of the form 'Fa v Ga'. In the first case the repeated letter '𝔔' represents the repetition of a quantifier 'P ⊃ ∼(∀x)(— x —)', and in the second case the repetition of a quantifier '(— a —)'.

[7] Recall that '(∃𝔮)(𝔮x)(Fx)' is an abbreviation for 'There is some true proposition of the form (with respect to "𝔔") "(𝔔x)(Fx)"'. Indeed, since every

too will contain a predicate somewhere, as is revealed by the fact that its form is represented with the help of a predicate-letter, but such propositions will not fall within our previous class of propositions.

We shall not overcome this problem by temporarily introducing a Russellian distinction into *orders*. We might perhaps say that all the formulae which we have explained already are *first-order* formulae, and that a proposition is a *first-order* proposition of the form '$(\mathfrak{Q}x)(Fx)$' if and only if it may be analysed as of a form represented by a first-order formula containing a predicate-letter somewhere. Introducing the formula '$(\mathfrak{Q}^1x)(Fx)$' and its alphabetic variants to represent the forms of these first-order propositions, we may then construct further formulae by combining these new formulae truth-functionally (with each other, and with all formulae hitherto available) and by applying universal quantification, and the further formulae thus formed may be called *second-order* formulae. A proposition may now be said to be a *second-order* proposition of the form '$(\mathfrak{Q}x)(Fx)$' if and only if it may be analysed as of a form represented by a second-order formula containing a predicate-letter somewhere, and we may introduce the formula '$(\mathfrak{Q}^2x)(Fx)$' to represent the form of all these propositions. Thus, a proposition of the form '$(\mathfrak{Q}^2x)(Fx)$' will typically involve a quantification over all propositions of the form '$(\mathfrak{Q}^1x)(Fx)$'. Proceeding in the same way we may evidently introduce propositions of the form '$(\mathfrak{Q}^nx)(Fx)$' for any finite number n, and we might then allow ourselves to introduce yet further formulae for propositions involving quantification over all propositions of any of the forms '$(\mathfrak{Q}^nx)(Fx)$' already available, and then indeed still further formulae for propositions involving quantification over these last propositions, and so on. At every stage we shall find new propositions of the form '$(\mathfrak{Q}x)(Fx)$', but at no stage will we have included them all, for indeed we shall not at any stage reach the example from which this discussion started, namely the form '$(\exists \mathfrak{q})(\mathfrak{q}x)(Fx)$'.

The reason for this is that any such hierarchical construction as we have been considering will have the property that a

proposition of such a form as '$(\forall x)(Fx \vee \sim Fx)$' is both true and of the form '$(\mathfrak{Q}x)(Fx)$', it is clear that every proposition of the form '$(\exists \mathfrak{q})(\mathfrak{q}x)(Fx)$' is in fact true.

proposition involving quantification over propositions of any form in the hierarchy will not itself be of that same form, but on the other hand we have just seen that a proposition of the form '$(\exists\mathfrak{q})(\mathfrak{q}x)(Fx)$' is a proposition which involves quantification over all propositions of the form '$(\mathfrak{Q}x)(Fx)$' and is itself of the form '$(\mathfrak{Q}x)(Fx)$', just as any proposition of the form '$(\exists\mathfrak{f})(\mathfrak{f}a)$' is itself of the form 'Fa'. With this last example we have avoided a similar difficulty only because we assumed an understanding of the notion of containing a subject somewhere in the first place, and it is of course quite obvious that a proposition of the form '$(\exists\mathfrak{f})(\mathfrak{f}a)$' does contain a subject somewhere. And the moral of the present argument is that in this case too we must in the end recognize that the notion of containing a predicate somewhere cannot be analysed in terms of the notions already available.

Actually, we could have reached this same conclusion on quite different grounds anyway. For our idea was that for a proposition to contain a predicate *somewhere* it must be built up in some way from a proposition which contains a predicate *initially*, and so far as it goes that idea is perfectly correct. But we then tried to specify the methods by which one proposition can be built up from others, and at this point we effectively limited our consideration to the methods of truthfunctional combination and universal quantification. But there seems no reason to suppose that these are the only available methods.

I have already touched upon one of the difficulties presented by such intensional propositions as those concerning belief, but even without entering this troubled area it does at least seem clear that we shall have to admit the existence of intensional (i.e. non-truthfunctional) propositional connectives, such as 'because'. Now a proposition which we might reasonably represent as of the form 'Fa because Ga' will be admitted by our previous account as a proposition which contains a predicate somewhere, just because it contains a predicate *initially* (viz: the predicate 'F... because G...'). But if we have two propositions which are of the related forms 'Fa because Ga' and 'Fa because Ha', then it seems quite clear that the first contains the predicate 'G...' wherever and only where the second contains the predicate 'H...', and consequently that the two propositions are of the related forms '$(\mathfrak{Q}x)(Gx)$' and '$(\mathfrak{Q}x)(Hx)$'. With our present limitation, however, the positions in which 'G...' and

'H ...' occur are apparently not recognized as predicate-positions at all. I suppose one might conceivably undertake to show that the connective 'because' can after all be analysed in purely extensional terms—involving presumably material implication, universal quantification, and no doubt some reference to times—but such an analysis would hardly be convincing just because the positions of 'G ...' and 'H ...' here seem quite clearly *not* extensional. We could indeed so restrict the notion of a subject-quantifier that a proposition was only to count as containing a predicate somewhere if it contained that predicate in an extensional position, and this would entitle us to adopt an axiom of extensionality for our subject-quantifiers, viz.

$$(\forall x)(Fx \equiv Gx) \supset ((\mathfrak{Q}x)(Fx) \equiv (\mathfrak{Q}x)(Gx)).$$

But it seems to me that this is a highly artificial restriction, and since we shall not find the axiom necessary for our purposes I prefer not to adopt it. Thus I recognize that there are probably other kinds of propositional connective than those that can be analysed in terms of the familiar truthfunctional connectives and universal quantification.

Similarly it seems to me that there may perhaps be other kinds of quantification that we would have to recognize here even if we were restricting our attention to extensional methods of building up propositions. Imagine, for instance, a community in which children are not taught any number-series but are taught to recognize groups of, say, five objects ostensively by their immediate appearance (i.e. without counting). The language of such a community might well contain an expression corresponding to our expression 'there are five . . .'—let us suppose it is the expression 'there are V . . .'—and in such a case I should want to say that propositions expressed by means of it were of the form '(There are Vx)(Fx)', and hence that they did contain a predicate and were of the form '$(\mathfrak{Q}x)(Fx)$'. But in the situation we are imagining it hardly seems satisfactory to say that these propositions are built up by universal quantification and truth-functional combination from propositions of the form 'Fa' (and, presumably, '$a \neq b$'), for the quantifier 'There are V. . .' is supposed to be entirely simple in meaning and not further analysable. (Unlike our quantifiers, we cannot say that its meaning is given by its position in a series, for the supposed community has no such series.) And even if this example seems unconvincing, I think we should still entertain some doubt over the sufficiency of the two methods of universal quantification and truth-functional combination, even where it is only extensional methods of construction that we are concerned with.

Both of these last points bear upon the adequacy of our account of *first-order* propositions of the form '$(\mathfrak{Q}x)(Fx)$', and it now appears that we cannot even regard these as a neat totality already accounted for. It may also be noted that just as we next found it necessary to include formulae such as '$(\exists\mathfrak{q})(\mathfrak{q}x)(Fx)$' which become available when the form '$(\mathfrak{Q}x)(Fx)$' becomes available, so we shall shortly find that yet further formulae containing predicate-letters become available as we ascend yet higher up the hierarchy of types, and these will not have been catered for in our original account. So on all these grounds it seems quite clear that we cannot hope to specify seriatim all the contexts in which predicates (or predicate-letters) may occur, and therefore that we cannot give a satisfactory account of the notion of containing a predicate somewhere in the way first envisaged.

I do not think that there is in fact any difficulty in understanding this notion, so long as we do understand the basic notion of containing a predicate *initially*, for the only further requirement is that we should see how propositions may be built up from component propositions which contain predicates initially. But this latter idea must be recognized as a primitive idea from a formal point of view, to be understood by projecting from the numerous examples already given, but not definable in terms of them. Similarly, I do not think that there is in fact any difficulty in understanding the idea of containing the same predicate somewhere, so long as we already understand what it is to contain the same predicate initially; and I do not think that there is in fact any difficulty in understanding the idea of one proposition's containing some predicate wherever and only where another contains some predicate, for this just is the idea that the two propositions are built up in the same way from two component propositions which each contain a predicate initially, this being what is tested by the substitution test I mentioned. The concept of a subject-quantifier, then, does not seem to me to be a difficult concept to grasp in the light of the explanations that I have been offering—indeed it seems in many ways much easier than the fundamental concepts of a subject and of a predicate—but in the end it, no less than they, must be admitted to be not formally definable.

I end this section, however, by stating a criterion for containing a subject-quantifier (initially) which is logically adequate and which is of some assistance despite its circularity. For it is perfectly true that a proposition contains a predicate somewhere, and so is of the form '($\mathfrak{Q}$x)(Fx)', if and only if it may also be analysed as of a form written with a predicate-letter somewhere *and no quantifier-letter*,[8] provided—and this is where the circularity enters—that we may employ quantifier-*variables* without infringing the ban on quantifier-*letters*. The explanations of containing the same predicate somewhere, and of containing the same subject-quantifier initially, which I have already given on pp. 87–8, may evidently be used in conjunction with this criterion.

2. The Hierarchy of Types

Before I come to discuss the general notion of a quantifier, which is my ultimate goal in this chapter, I shall pause to develop the hierarchy of types a little further, for this is of some relevance here. However, I confine my attention to what one might call the hierarchy of *monadic* types, quite ignoring the classification of relations of different levels, for that will be as far as we need to take the subject. And to begin with I shall adopt the traditional way of speaking by which it is various different (monadic) entities that are classified into types, namely subjects (or objects), predicates, subject-quantifiers, and so on (but soon it will become more convenient to change to a less misleading terminology). With this approach the hierarchy of monadic types can be very simply explained thus: the first type comprises subjects, the second the remainders to subjects (viz. predicates), the third the remainders to the remainders to subjects (viz. subject-quantifiers), the fourth the remainders to the remainders to the remainders to subjects, and so on. And we may explain what is meant by the notion of a *remainder* in this context by

[8] To avoid confusion at this point I should repeat that when I speak of a quantifier-letter I always mean a *schematic* quantifier letter, i.e. one of the letters '$\mathfrak{Q}$', '$\mathfrak{Q}_1$', '$\mathfrak{Q}_2$',... For example, '$\forall$' is *not* a quantifier letter.

saying that Φ-components are remainders to α-components if and only if every proposition which contains an α-component *somewhere* also contains a Φ-component *initially*, and so is of the elementary Φ-α form. Essentially this is the principle on which Frege would have proceeded had he extended his construction beyond the level of subject-quantifiers, and much of what Russell had to say about his theory of types suggests that he would not have dissented from it, so far as it goes. But there is one very notable respect in which this principle leads to results quite different from those assumed by Russell, and that is in the matter of *significance*. It will prevent misunderstanding if I point out this divergence at once.

In chapter 2, while expounding the distinction between subjects (or objects) and predicates, I remarked that the combination of symbols 'F(F)' was not significant, since a predicate-letter has a use only in conjunction with a subject-letter or a subject-variable; and we could equally add that the combination of symbols 'a(a)' is not significant, since a subject-letter has a use only in conjunction with a predicate-letter or a predicate-variable. Conversely elementary subject-predicate formulae such as 'Fa' are significant—i.e. there are propositions of that form—and so therefore are any combinations of symbols built up from these elementary formulae by truth functional combination and quantification. From these principles of significance all other principles that affect us may be derived.

First it is worth noting that for our entities of third type, namely subject-quantifiers, we have adopted the notation '($\mathfrak{Q}$x)(—x—)' which contains a subject-variable as part of itself, and so when we construct the elementary formulae involving symbols of this type by filling the gap with a predicate-letter, as in '($\mathfrak{Q}$x)(Fx)', we do not offend the principle that a predicate-letter can occur only in conjunction with a subject-letter or a subject-variable. In fact the only significant way of filling the gap in the schema '($\mathfrak{Q}$x)(—x—)' so as to yield a complete formula is with a predicate-letter or more generally with a predicate-*schema* (as e.g. in '($\mathfrak{Q}$x)(Fx & Gx)'). And the only significant way of filling this gap so as to yield a part of a formula is by first introducing a predicate-schema and then perhaps exchanging some of the schematic letters in that schema for their associated variables (and adding initial quantifiers to bind them, as in

'(∀𝔣)(...(𝔔x)(𝔣x)...)'). Now I stated in my discussion of the universal quantifier in chapter 3 that a formula such as '(∀𝔣)(...(𝔔x)(𝔣x)...)' is significant if and only if its part '(...(𝔔x)(Fx)...)' is significant[9] and so the distinction between schematic letters and variables is of no importance in this connection. Ignoring the distinction, then, we may say simply that a subject-quantifier schema has a use only in conjunction with a predicate-schema, and this of course is precisely the range of situations in which a subject-letter has a use. So we may say that subjects and subject-quantifiers have precisely the same range of significance, in the sense that we shall always preserve significance by substituting a subject-quantifier expression for a subject-expression, and conversely. Furthermore this conclusion concerning subjects and subject-quantifiers in fact holds quite generally for any pair of propositional components of our hierarchy such that the one is a remainder to a remainder to the other, so that types which are next but one to one another will always have the same range of significance. Consequently the range of significance of our first type, subjects, will also be the range of significance of all odd-numbered types, and the range of significance of our second type, predicates, will also be the range of significance of all even-numbered types. Generally, a significant formula will be obtained by coupling any expression of any even-numbered type with any expression of any odd-numbered type, but in no other way.

Perhaps it will help if I now develop the hierarchy explicitly a little further. As entities of our third type we have the remainders to the remainders to subjects, which we called subject-quantifiers, and as entities of our fourth type we have the remainders to the remainders to predicates, which I propose to call predicate-quantifiers. To see the rationale for this, let us first consider that if we start with any proposition which contains a subject-quantifier initially, and so is of the form '(𝔔x)(Fx)', then deleting the subject-quantifier from that proposition will leave us what is here represented by a predicate-letter surrounded by a gap, thus '(—F—)'; hence each predicate will have its associated predicate-quantifier just as each subject has its associated subject-quantifier. Again, if we now consider propositions which contain subject-quantifiers but not initially, one of the simplest

[9] See pp. 79–81.

cases seems to be the case in which the subject-quantifier is completed not with a predicate but with a predicate-variable, as for instance in the proposition '$(\exists\mathfrak{f})(\forall x)(\mathfrak{f}x)$'; but if we delete the subject-quantifier from this proposition then what we have left will be represented by the schema '$(\exists\mathfrak{f})(—\mathfrak{f}—)$', and this is a paradigm expression for a predicate-quantifier. Of course predicate-quantifiers, as they are here defined, will go beyond these paradigm cases, but they will do so in ways precisely analogous to the ways in which subject-quantifiers go beyond *their* paradigm cases, so there is no new ground for dissatisfaction here.

By analogy with the case of subject-quantifiers I propose to adopt the notation '$(\mathfrak{Q}\mathfrak{f})(—\mathfrak{f}—)$' as my expression for an elementary predicate-quantifier schema, and I say that this may be completed with an elementary subject-quantifier schema '$(\mathfrak{Q}_1 x)(—x—)$' to yield the elementary formula '$(\mathfrak{Q}\mathfrak{f})(\mathfrak{Q}_1 x)(\mathfrak{f}x)$'. Indeed it will now be more convenient if we attend not so much to the supposed 'entities' predicates, subject-quantifiers, predicate-quantifiers, and so on, but consider rather the elementary formulae associated with them. So far I have introduced three of these, namely

(i) Fa

(ii) $(\mathfrak{Q}_1 x)(Fx)$

(iii) $(\mathfrak{Q}_2 \mathfrak{f})(\mathfrak{Q}_1 x)(\mathfrak{f}x)$.

I shall say of these formulae, and their successors, that they are the elementary formulae of the orthodox hierarchy of (monadic) types, and that each is the *characteristic* formula of a type. Types may now be regarded as classifications of formulae (and hence of propositional forms, and therefore ultimately of propositions), and I shall say that a formula is of type n if and only if it can be obtained by legitimate substitution from the characteristic formula of type n, the characteristic formulae of the first three types being those displayed above. But how are we now to go on to construct further characteristic formulae?

Well, from a heuristic point of view we have now to consider the remainders to predicate-quantifiers, i.e. what is left when a predicate-quantifier is deleted from a proposition. If we take any proposition containing a predicate-quantifier initially and delete that quantifier we are left evidently with a subject-quantifier surrounded by a gap, thus '$(...(\mathfrak{Q}x)(—x—)...)$'; hence each subject-quantifier again has its associated member of the type next above predicate-quantifiers. Again if we look for the simplest type of proposition which contains a predicate-quantifier but not initially, we will light upon an example such as '$(\forall\mathfrak{q})(\exists\mathfrak{f})(\mathfrak{q}x)(\mathfrak{f}x)$', and deleting the predicate-quantifier from this will yield what is represented by the schema

$$\text{'}(\forall\mathfrak{q})(...(\mathfrak{q}x)(—x—)...)\text{'},$$

which is plainly the expression for a *subject-quantifier*-quantifier. So we may say that the next type above predicate-quantifiers is that of subject-quantifier-quantifiers, and I accordingly adopt the notation '$(\mathfrak{Q}_1\,\mathfrak{q})(...(\mathfrak{q}x)(—x—)...)$' as that of an elementary subject-quantifier-quantifier schema, which I say may be completed with an elementary predicate-quantifier schema '$(\mathfrak{Q}_2\,\mathfrak{f})(—\mathfrak{f}—)$' to yield the elementary formula

$$\text{'}(\mathfrak{Q}_1\,\mathfrak{q})(\mathfrak{Q}_2\,\mathfrak{f})(\mathfrak{q}x)(\mathfrak{f}x)\text{'}.$$

Less misleadingly we may say that this last formula is the *characteristic* formula of the fourth type, and by the same principle we may construct the characteristic formula of the fifth type by taking the elementary schema for predicate-quantifier-quantifiers and completing it with the elementary schema for subject-quantifier-quantifiers, and so on. In fact the characteristic formulae of the orthodox hierarchy of (monadic) types are just the formulae in the series:

Fa

$(\mathfrak{Q}_1\,x)(Fx)$

$(\mathfrak{Q}_2\,\mathfrak{f})(\mathfrak{Q}_1\,x)(\mathfrak{f}x)$

$(\mathfrak{Q}_3\,\mathfrak{q}_1)(\mathfrak{Q}_2\,\mathfrak{f})(\mathfrak{q}_1\,x)(\mathfrak{f}x)$

$(\mathfrak{Q}_4\,\mathfrak{q}_2)(\mathfrak{Q}_3\,\mathfrak{q}_1)(\mathfrak{q}_2\,\mathfrak{f})(\mathfrak{q}_1\,x)(\mathfrak{f}x)$

$(\mathfrak{Q}_5\,\mathfrak{q}_3)(\mathfrak{Q}_4\,\mathfrak{q}_2)(\mathfrak{q}_3\,\mathfrak{q}_1)(\mathfrak{q}_2\,\mathfrak{f})(\mathfrak{q}_1\,x)(\mathfrak{f}x)$

etc.

Now I will certainly admit that this notation is rather cumbersome, and on any occasion on which we intend to put these formulae to serious use it will be much more convenient to adopt something simpler, but the present notation is I think a perspicuous one from a philosophical point of view. First, it shows very clearly how all succeeding formulae are built up step by step from the initial subject-predicate formulae, and so makes it easier to see why the principles of significance I have stated do indeed obtain. Once we have grasped that a schematic expression which can significantly be completed with a subject-letter 'a' can also be significantly completed with a subject-quantifier-schema '$(\mathfrak{Q}x)(\text{—}x\text{—})$', and conversely, it is not difficult to see that the same must hold also for subject-quantifier schemata and the schemata obtained from them in turn by a step of quantification, namely schemata such as '$(\mathfrak{Q}\mathfrak{q})(\ldots(\mathfrak{q}x)(\text{—}x\text{—})\ldots)$'. For an expression of the form '$(\mathfrak{Q}\mathfrak{q})(\ldots(\mathfrak{q}x)(\text{—}x\text{—})\ldots)$' will be significant if and only if its part '$(\ldots(\mathfrak{Q}x)(\text{—}x\text{—})\ldots)$' is significant, and this as we have seen will be significant if and only if the corresponding expression '$(\ldots(\text{—}a\text{—})\ldots)$' is significant, which is precisely the point to be demonstrated. And from the way in which the hierarchy of characteristic formulae is constructed it is clear that exactly the same argument will be repeatable at any stage in the construction.[10]

Another revealing feature of the notation is that the whole hierarchy of formulae is built up with only three different alphabets of schematic letters (and associated variables), namely subject-letters, predicate-letters, and quantifier-letters. What provides for the ever-increasing complexity of the formulae is not the continual addition of new alphabets of schematic letters, but ever more complicated arrangements of the quantifier-letters. In fact all the characteristic formulae of the hierarchy are obtained simply by prefixing strings of quantifiers to the initial subject predicate formulae. At this point one might naturally raise the question whether we have included *all* formulae that can be obtained in this way; I have for instance made no mention of formulae such as

$$\text{'}(\mathfrak{Q}\mathfrak{f})(\mathfrak{f}a)\text{'} \qquad \text{or} \qquad \text{'}(\mathfrak{Q}_3\,\mathfrak{q}_1)(\mathfrak{q}_1\,x)(\mathfrak{Q}_2\,\mathfrak{f})(\mathfrak{f}x)\text{'}$$

yet it seems clear that these ought to be included in the hierarchy somewhere. However, it turns out that in the relevant sense all these formulae *are* already included, since the ones I have not mentioned explicitly are all special cases of the ones I have mentioned. (And when I say that one formula is a special case of

[10] A consequence is that schemata such as '$(\mathfrak{Q}_1x)(\text{—}x\text{—})$' and '$(\mathfrak{Q}_2\,\mathfrak{f})(\text{—}\mathfrak{f}\text{—})$' of adjacent types may be coupled in two ways to yield a significant formula, either as '$(\mathfrak{Q}_1\,x)(\mathfrak{Q}_2\,\mathfrak{f})(\mathfrak{f}x)$' or as '$(\mathfrak{Q}_2\,\mathfrak{f})(\mathfrak{Q}_1\,x)(\mathfrak{f}x)$'. There can be no analogue to this on Russellian principles.

another I mean that the one can be obtained from the other by a legitimate application of the rule of substitution, and hence that all propositions of the one form are thereby propositions of the other form also.)

To treat of this point economically, let us temporarily introduce a briefer symbolism. In place of the elementary subject predicate formula 'Fa' let us now write 'PS', where 'P' represents the predicate-component and 'S' the subject-component. Now the formulae which can be constructed from 'Fa' by prefixing just one quantifier-expression are '$(\mathfrak{Q}x)(Fx)$' and '$(\mathfrak{Q}\mathfrak{f})(\mathfrak{f}a)$', and in place of these let us now write 'SPS' and 'PPS' respectively. Again, the formulae which can be constructed from these in turn by prefixing just one quantifier-expression are '$(\mathfrak{Q}_2\, \mathfrak{q}_1)(\mathfrak{q}_1\, x)(Fx)$', '$(\mathfrak{Q}_2\, \mathfrak{f})(\mathfrak{Q}_1\, x)(\mathfrak{f}x)$', '$(\mathfrak{Q}_2\, x)(\mathfrak{Q}_1\, \mathfrak{f})(\mathfrak{f}x)$' and '$(\mathfrak{Q}_2\, \mathfrak{q}_1)(\mathfrak{q}_1\, \mathfrak{f})(\mathfrak{f}a)$', and these we will now write as 'SSPS', 'PSPS', 'SPPS', 'PPPS', respectively. And so on. Roughly, the idea is that subjects, subject-quantifiers, subject-quantifier-quantifiers, and so on, are all replaced by 'S', while predicates, predicate-quantifiers, predicate-quantifier-quantifiers, and so on, are all replaced by 'P'. In this way any formula of the sort we are inquiring into will be replaced by a sequence of 'P' 's and 'S' 's, in which the last letters are 'PS', and in which the first 'P' and the first 'S' represent a schematic letter for a quantifier which binds the variable represented by the next 'P' and the next 'S', which in turn binds the variable represented by the next 'P' and the next 'S', until eventually we come to the predicate-variable (or letter) represented by the last 'P' and the subject-variable (or letter) represented by the last 'S'.

Now when formulae are written in this way, the rules governing the assignment of formulae to types may be stated very simply thus:

(i) The formula 'PS' is of type 1.

(ii)*a* Prefixing 'P' to any formula already beginning with 'P' leaves the type of the formula unchanged.

(ii)*b* Prefixing 'S' to any formula already beginning with 'S' leaves the type of the formula unchanged.

(iii)*a* Prefixing 'P' to any formula previously beginning with 'S' increases the type of the formula by one.

(iii)*b* Prefixing 'S' to any formula previously beginning with 'P' increases the type of the formula by one.

Applying these rules we may then sort all the relevant formulae into types thus:

Type 1	*Type* 2	*Type* 3	*Type* 4	etc.
PS				
PPS	SPS			
PPPS	SPPS	PSPS		
	SSPS			
PPPPS	SPPPS	PSPPS	SPSPS	
	SSPPS	PSSPS		
	SSSPS	PPSPS		
etc.				

Evidently all formulae may be sorted on this principle, and we find that formulae built up from just n letters will have a binomial distribution amongst the first n−1 types, while the formula of n letters which is of the highest type—in fact of the (n−1)'th type—is precisely the formula which I have listed as the characteristic formula of that type. These characteristic formulae each begin the list of formulae of that type, for they have as special cases all the formulae listed below them.

So far I have simply stated the rules for determining the type of a formula, but not explained why they are correct. However, this is in fact very easy to see, for rules (i) and (iii) simply reflect our initial definition of the hierarchy of types, while rule (ii) is an immediate result of our way of introducing the characteristic formulae. Thus it is clear that any proposition of the form '$(\mathfrak{Q}\mathfrak{f})(\mathfrak{f}a)$' will be a proposition which contains a subject (since its form is here represented with the help of a subject-letter), and consequently it will also be of the form 'Fa'; in other words, the formula 'PPS' is a special case of the formula 'PS', and this justifies the second stage in our construction of the list. From this in turn it now follows that 'P(PPS)' and 'S(PPS)' are special cases of 'P(PS)' and 'S(PS)' respectively, which justifies the placing of all formulae in the third stage except one. And further, it is clear that any proposition of the form '$(\mathfrak{Q}_2\mathfrak{q}_1)(\mathfrak{q}_1 x)(Fx)$' contains a predicate somewhere, and is therefore of the form '$(\mathfrak{Q}x)(Fx)$', which shows that the remaining formula 'SSPS' is a special case of 'SPS' and completes the justification of the third stage of the list. By elaborating these results as before we shall justify the placing of all formulae in the fourth stage except one, namely 'PPSPS', and this will then be justified by the account of what it is for a proposition to be of the form 'PSPS'. By continuing this procedure we shall evidently be able to justify all applications of rule (ii).

Finally, I should point out that each of the characteristic formulae is a special case of the characteristic formula of next higher type, and therefore of the characteristic formulae of all higher types; thus a proposition of the form 'Fa' is also of the form '$(\mathfrak{Q}x)(Fx)$' just because it may always be viewed as got by applying the subject-quantifier '(—a—)' to the predicate 'F...', and similarly a proposition of the form '$(\mathfrak{Q}_1 x)(Fx)$' is also of the form '$(\mathfrak{Q}_2 \mathfrak{f})(\mathfrak{Q}_1 x)(\mathfrak{f}x)$' because again it may equally well be viewed as got by applying the predicate-quantifier '(—F—)' to the subject-quantifier '$(\mathfrak{Q}_1 x)(-x-)$'. And so on. (In fact a subject–predicate proposition is also of *every one* of the forms we have been considering). The converse, however, is not true—and if it were there would be no point in making these distinctions of type. Thus a proposition such as '$(\forall x)(x = x)$' is of the form '$(\mathfrak{Q}x)(Fx)$' but is not of the form 'Fa', a proposition such as '$(\forall \mathfrak{f})(\exists x)(\mathfrak{f}x)$' is of the form '$(\mathfrak{Q}_2 \mathfrak{f})(\mathfrak{Q}_1 x)(\mathfrak{f}x)$' but is not of the form '$(\mathfrak{Q}x)(Fx)$', and so on.

I think that this discussion suffices to show how all the formulae we have been considering may be sorted into types, and what are the relations between these types, and so we may now return to a further consideration of the explanation of these formulae. For the explanations I have given or suggested so far require supplementing in one important respect.

I have claimed it as a merit of the notation I adopt that it requires only three distinct alphabets of schematic letters, namely those for subjects, for predicates, and for quantifiers. In this notation the same schematic quantifier-letters occur over and over again in ever more complicated contexts, and the reason why I have used the same schematic letters in all these various contexts is that there are some constants, for instance '$\forall$' or '$\sim(\forall ...)\sim$', which may be substituted for those letters in all their contexts without loss of significance. This is because, just as any formula of the orthodox hierarchy can be negated, so also any such formula can be universally quantified, and in each case it is recognizably the same operation that is performed, whatever the type of the formula we are dealing with. (That negation, and other truthfunctional operations, apply unchanged in any type is perfectly evident; that the same holds for universal quantification was the burden of my argument in chapter 3.) This fact is important to our aim of constructing arithmetic, for as I pointed out earlier in chapter 1 (pp. 19–20) that construction will require us to introduce numerical quantifiers in such a way that they too can be seen to be applicable in any type, just as the universal quantifier can.

But when we reflect upon the explanation I have so far given of the formulae of the orthodox hierarchy, we find that this important feature of the notation is so far not accounted for. For I began by discussing the explanation of subject-quantifiers in some detail, and in particular the notion of containing the same subject-quantifier. Since then I have assumed that analogous explanations are to be given of predicate-quantifiers, subject-quantifier-quantifiers, and so on, so that we now understand the notions of containing the same predicate-quantifier and containing the same subject-quantifier-quantifier. But in order to explain the significance of repeating the same quantifier-letter in formulae such as '$(\mathfrak{Q}x)(Fx)$' and '$(\mathfrak{Q}\mathfrak{f})(\mathfrak{Q}_1 x)(\mathfrak{f}x)$', where the quantifier-letter binds variables of different types, it seems that we must now offer a new and more general explanation of the notion simply of *containing the same quantifier*, and this will apparently take us beyond those special cases of quantifiers that we have so far been considering. However, it turns out that for the immediate purpose of completing the explanation of the formulae of the orthodox hierarchy of types no radically new idea is required.

It will be helpful at this point if I first recapitulate the explanation previously given, but applying it to a type of case that I have not so far considered explicitly (though it is covered by the previous explanation), namely a case where the same quantifier binds a variable of the same type but in formulae of different types. An example would be what is signified by the repetition of the same quantifier-letter in the two formulae '$(\mathfrak{Q}\mathfrak{f})(\mathfrak{Q}_1 x)(\mathfrak{f}x)$' and '$(\mathfrak{Q}\mathfrak{f})(\mathfrak{f}a)$'. Since predicate-quantifiers are to be understood in the same way as subject-quantifiers, we are to understand that a proposition is of the form '$(\mathfrak{Q}\mathfrak{f})(\mathfrak{Q}_1 x)(\mathfrak{f}x)$' if and only if its form can also be represented by a formula containing the schema '$(\mathfrak{Q}_1 x)(—x—)$' somewhere (and containing no predicate-quantifier-letter), and similarly a proposition is of the form '$(\mathfrak{Q}\mathfrak{f})(\mathfrak{f}a)$' if and only if its form can also be represented by a formula containing the letter 'a' somewhere (and containing no predicate-quantifier-letter). So it is further to be understood that two propositions are of these related forms if and only if the relevant formulae representing their related forms differ from one another only in that the one contains the schema '$(\mathfrak{Q}_1 x)(—x—)$' wherever and only where the other contains the letter 'a', and this in turn will be so if and only if the first can be obtained from the second by substituting the schema '$(\mathfrak{Q}_1 x)(—x—)$' uniformly for every occurrence of 'a' in the second, and the second can be obtained from the first by the reverse substitution.

The last clause of this explanation is satisfactory as it stands only because we can significantly substitute a subject-quantifier-schema for a subject-letter, and conversely, but it will not be satisfactory where we have started with such a pair of formulae as '$(\mathfrak{Q}x)(Fx)$' and '$(\mathfrak{Q}\mathfrak{f})(\mathfrak{Q}_1 x)(\mathfrak{f}x)$'. For the result of substituting the letter 'F' for the schema '$(\mathfrak{Q}_1 x)(—x—)$' will always be a non-significant string of symbols, since a predicate-letter must be completed with a subject-letter or a subject-variable while a subject-quantifier-schema cannot be completed in this way. The difficulty, however, can be easily overcome if we slightly modify the explanation of the condition that the one formula should differ from the other only in that the one contains '$(\mathfrak{Q}_1 x)(—x—)$' wherever and only where the other contains 'F', by adding that when the one schema is substituted for the other it is permissible at the same time to substitute for any *bound variable* completing the one the appropriate bound variable completing the other, if this is necessary to preserve significance. So for instance the pairs of formulae

$(\forall x)(Fx)$	and	$(\forall \mathfrak{f})(\mathfrak{Q}_1 x)(\mathfrak{f}x)$
$\sim(\forall x)(\sim Fx)$	and	$\sim(\forall \mathfrak{f})(\sim(\mathfrak{Q}_1 x)(\mathfrak{f}x))$
$(\forall x)(Fx \mathbin{\&} \sim Fx)$	and	$(\forall \mathfrak{f})((\mathfrak{Q}_1 x)(\mathfrak{f}x) \mathbin{\&} \sim(\mathfrak{Q}_1 x)(\mathfrak{f}x))$
$P \mathbin{\&} (\forall x)(Fx)$	and	$P \mathbin{\&} (\forall \mathfrak{f})(\mathfrak{Q}_1 x)(\mathfrak{f}x)$
$(\forall x)(P \mathbin{\&} Fx)$	and	$(\forall \mathfrak{f})(P \mathbin{\&} (\mathfrak{Q}_1 x)(\mathfrak{f}x))$

will now be held to satisfy the condition that the one differs from the other only in that the one contains '$(\mathfrak{Q}_1 x)(-x-)$' wherever and only where the other contains 'F', and accordingly propositions which are of these pairs of related forms may also be said to be of the related forms '$(\mathfrak{Q}x)(Fx)$' and '$(\mathfrak{Q}\mathfrak{f})(\mathfrak{Q}_1 x)(\mathfrak{f}x)$'. Evidently this revised criterion for containing the same quantifier (initially) is entirely faithful to our original intention, for the pairs of formulae envisaged can certainly be regarded as *built up in the same way* from the different starting points 'F...' and '$(\mathfrak{Q}_1 x)(-x-)$' by the same steps of truthfunctional combination and quantification. (And if we are adopting the restrictive view canvassed in section 1 by which the only ways of building up a formula are by steps of truthfunctional combination and universal quantification, then the effect of this criterion is just to draw attention to the fact that these are both 'type-neutral' operations.)

But we should observe here that even with this understanding of being built up in the same way, we cannot say that for any formula that is built up in some way from one propositional component there will always be another formula built up in the same way from any other propositional component. For it will not always be possible to perform the required substitution, even as now understood, while preserving significance. For example we cannot substitute 'F' for '$(\mathfrak{Q}_1 x)(-x-)$' in either of the formulae

$$(\mathfrak{Q}_1 x)(Fx)$$

$$(\forall \mathfrak{f})((\mathfrak{Q}_1 x)(\mathfrak{f}x) \vee (\mathfrak{Q}_2 x)(\mathfrak{f}x))$$

without losing significance, and therefore although propositions of either of these forms will also be of the form '$(\mathfrak{Q}\mathfrak{f})(\mathfrak{Q}_1 x)(\mathfrak{f}x)$', there will not be any propositions related to these as of the related form '$(\mathfrak{Q}x)(Fx)$'. In these examples the quantifier introduced by '$(\mathfrak{Q}\mathfrak{f})(-\mathfrak{f}-)$' is unavoidably a *predicate*-quantifier, and it is not possible to apply that same quantifier where one could apply a subject-quantifier.

This completes the explanation of the formulae of the orthodox hierarchy of monadic types—or, otherwise put, the explanation of the types of the orthodox hierarchy whose characteristic formulae contain just two schematic letters. It is not difficult to see how the account might be extended to cover polyadic propositional components generally, but for our purposes there is no need to develop this extension, for the quantifiers we are mainly interested in are the numerical quantifiers, and these are monadic. Rather, we should now look further into the point just made in the last paragraph, that although there are some cases in which the same quantifier may be applied unchanged to any schema of the hierarchy, there are also others in which the quantifier can only be applied to schemata of certain types. In the latter case the type of the variable bound by the quantifier is as it were part of the nature of the quantifier itself, while in the former this is not so; and this point evidently calls for further discussion.

3. The Concept of a Quantifier

In ordinary discourse a quantifier very often occurs in connection with what we may easily view as a 'restriction' on that quantifier. Thus when we are dealing with a phrase of the form 'All of the so-and-so's' (as in 'All of the men in this room are wearing shoes') I say that the part '. . . of the so-and-so's' expresses a *restriction* on the quantifier expressed by 'all', and where we are dealing with a phrase of the form 'All so-and-so's' (as in 'All men are mortal') I say that it is the part '. . . so-and-so's' that expresses the restriction. However, a proposition containing a quantifier that is restricted in this fashion can always be paraphrased in such a way that the quantifier appears quite unrestricted, and this is in a sense our standard practice in logic. When

All men are mortal
Some men are mortal
Only men are mortal

are paraphrased to

$$(\forall x)(x \text{ is a man} \supset x \text{ is mortal})$$
$$(\exists x)(x \text{ is a man} \;\&\; x \text{ is mortal})$$
$$(\forall x)(x \text{ is mortal} \supset x \text{ is a man})$$

the restricted quantifiers previously expressed by 'all men', 'some men', and 'only men', are still expressed by the quantifier-expressions

$$(\forall x)(x \text{ is a man} \supset \text{—}x\text{—})$$
$$(\exists x)(x \text{ is a man} \;\&\; \text{—}x\text{—})$$
$$(\forall x)(\text{—}x\text{—} \supset x \text{ is a man})$$

but the propositions may also be seen not as containing these quantifiers but as containing the unrestricted universal and existential quantifiers expressed simply by '∀' and '∃'. One might be inclined to say that the quantifiers they contain are expressed rather by the schemata '(∀x)(—x—)' and '(∃x)(—x—)', and these are still expressions for *restricted* quantifiers, namely quantifiers restricted to *subjects*. But it seems to me that both ways of viewing the matter are equally permissible, and the first is the more important to our present approach.

Earlier I made the point that a quantifier can always be seen as stating of a propositional form that certain of the propositions of that form are true, and in the proposition 'all men are mortal' we can certainly regard the restricted quantifier 'all men' as being used to state of the form 'a is mortal' that all propositions of that form *which have men as their subjects* are true. But when this proposition is rephrased as '(∀x)(x is a man ⊃ x is mortal)' it seems clear that we are being invited to see it equally as obtained by applying a quantifier to the form '(a is a man ⊃ a is mortal)', and it is evidently quite unnecessary to take the quantifier as stating that all propositions of that form *which have subjects* are true, since the form itself is already written with a schematic subject-letter, which shows that all propositions of that form must have subjects anyway. I do not want to say that we cannot view the proposition as containing a quantifier restricted to subjects—for we certainly can—but only that we can *also* view it as containing an unrestricted quantifier, since the restriction to subjects is already present anyway in what the quantifier is applied to.[11]

It is important here to observe that while the propositional *form* 'a is a man ⊃ a is mortal' carries a restriction to subjects within itself, the (one-place) *predicate* '... is a man ⊃ ... is mortal' has no such internal restriction; one may obtain a proposition from it by supplying it with a subject such as 'Socrates', but also by supplying it with a subject-quantifier such as 'All men', or in yet further ways. This may suggest that the subject-quantifiers,

[11] As Russell put it, the restriction imposed upon the quantified variable is 'a natural restriction upon the possible values of (i.e. a restriction given with) the function "if x is a man, x is mortal", and not needing to be imposed from without'. ('Mathematical Logic as based on the Theory of Types', in *Logic and Knowledge*, p. 71.)

predicate-quantifiers, and so on that we have hitherto been treating of should be sharply distinguished from the quantifiers proper that we are now considering, on the ground that the former yield propositions when applied to *predicates* and the like, while the latter yield propositions when applied to propositional *forms*. So one might say that when we are working within the orthodox hierarchy of types we view a formula such as '(∀x)(Fx)' as breaking into the two parts '(∀x)(—x—)' and 'F ...', while when we consider it as representing a quantifier applied to a propositional form we view it rather as breaking into the two parts '∀' and '(... x)(Fx)'—where the variable 'x' which forms part of '(... x)(Fx)' has (besides its usual job of indicating the bondage links) the further role of showing that the form in question is one written with a subject-letter. However, it seems to me that there is nothing wrong with combining the two approaches, so that we may in addition regard the formula as breaking into a subject-restricted quantifier '(∀x)(—x—)' and a propositional form '(... x)(Fx)'. In this way we continue to view the various restricted quantifiers of the orthodox hierarchy as still perfectly good quantifiers alongside the pure or unrestricted quantifiers such as '∀'; both sorts of quantifier are regarded as applied to propositional forms, and our logic of quantifiers would treat of them impartially, except for the fact that a restricted quantifier is applicable only to certain forms (namely those written with the appropriate schematic letter) while a pure quantifier is applicable to all forms without exception. There is perhaps some danger of confusion, for if we do not clearly distinguish between thinking of a proposition such as '(∃𝔣)(∀x)(𝔣x)' as formed by applying a predicate-quantifier to the subject-quantifier '(∀x)(—x—)' and to the propositional form '(∀x)(Fx)' (i.e. '(... 𝔣)(∀x)(𝔣x)'), we may unwarily fail to distinguish between the subject-quantifier and the propositional form. And if we are led on to suppose that even an unrestricted quantifier can also be regarded as a propositional form, we would have to admit that—since it is applicable to *any* monadic propositional form—it must be literally applicable to itself. But there is no need to become enmeshed in Russell's paradox in this way, and a logic which treats impartially of both pure and restricted quantifiers is certainly possible.

Nevertheless, I shall not in fact develop such a logic in any detail. My main reason is that for the purposes of constructing arithmetic it will not be necessary to employ a symbolism in which schematic quantifier-letters are taken to range over all sorts of quantifiers, both pure and restricted, while it will be necessary on occasion to limit our consideration to pure quantifiers,[12] and so we might as well introduce this limitation early on. A further consideration is that the more general logic must embody some unexpected restrictions on the application of the usual rules of *propositional* logic if we are not to fall into a fallacy. These restrictions stem from the fact (pointed out on p. 105) that although there are propositions of such related forms as '$(\mathfrak{Q}x)(Fx)$' and '$(\mathfrak{Q}\mathfrak{f})(\exists x)(\mathfrak{f}x)$'—namely propositions which contain a *pure* quantifier initially—and although any proposition of the form 'Fa' is also of the form '$(\mathfrak{Q}x)(Fx)$', nevertheless there are no propositions so related to it that it and they are of the related forms '$(\mathfrak{Q}x)(Fx)$' and '$(\mathfrak{Q}\mathfrak{f})(\exists x)(\mathfrak{f}x)$'. Ignoring this point may lead to fallacies, as we may verify by observing that in view of the trivial theorem

(i) $\vdash(\mathfrak{Q}\mathfrak{f})(\exists x)(\mathfrak{f}x) \vee \sim(\mathfrak{Q}\mathfrak{f})(\exists x)(\mathfrak{f}x)$

we expect to be able to deduce

(ii) $\vdash(\forall\mathfrak{f})((\mathfrak{Q}x)(\mathfrak{f}x) \equiv \mathfrak{f}a) \supset ((\forall\mathfrak{f})((\mathfrak{Q}x)(\mathfrak{f}x) \equiv \mathfrak{f}a)$
$\&((\mathfrak{Q}\mathfrak{f})(\exists x)(\mathfrak{f}x) \vee \sim(\mathfrak{Q}\mathfrak{f})(\exists x)(\mathfrak{f}x)))$

and from this in turn we expect it to follow that

if (iii) $\vdash\sim((\forall\mathfrak{f})((\mathfrak{Q}x)(\mathfrak{f}x) \equiv \mathfrak{f}a)$
$\& ((\mathfrak{Q}\mathfrak{f})(\exists x)(\mathfrak{f}x) \vee \sim(\mathfrak{Q}\mathfrak{f})(\exists x)(\mathfrak{f}x)))$

then (iv) $\vdash\sim(\forall\mathfrak{f})((\mathfrak{Q}x)(\mathfrak{f}x) \equiv \mathfrak{f}a)$.

However, under the intended interpretation (iii) is true and (iv) is false. There are indeed propositions of the form

$(\forall\mathfrak{f})((\mathfrak{Q}x)(\mathfrak{f}x) \equiv \mathfrak{f}a) \& ((\mathfrak{Q}\mathfrak{f})(\exists x)(\mathfrak{f}x) \vee \sim(\mathfrak{Q}\mathfrak{f})(\exists x)(\mathfrak{f}x))$,

for we may substitute any pure quantifier, say '$\exists$', in place of the schematic letter '$\mathfrak{Q}$' in this formula, and the result will be a proposition. However only *pure* quantifiers can significantly take the place of '$\mathfrak{Q}$' here, and for any *pure* quantifier in place of '$\mathfrak{Q}$' the resulting proposition is false, because its first conjunct is false (provided that there exists more than one object); hence

[12] See especially pp. 156–7.

(iii) is true. On the other hand (iv) is false, for there certainly are true propositions of the form

$$(\forall \mathfrak{f})((\mathfrak{Q}x)(\mathfrak{f}x) \equiv \mathfrak{f}a),$$

namely any proposition in which we take '$\mathfrak{Q}$' not as a pure quantifier but as the subject-restricted quantifier associated with the subject 'a'. This breakdown of the rules of ordinary propositional logic has not arisen, as on p. 80, through the non-existence of propositions of some of the forms in question, for we have seen that there are propositions of all of the forms (i)–(iv). The trouble is that the schematic letter '$\mathfrak{Q}$' can only be taken to stand in for pure quantifiers in formulae (ii) and (iii), since no other substitution would preserve significance, while in (iv) it *also* stands in for subject-restricted quantifiers; and that is what makes (iv) false while (ii) and (iii) are true. In a nutshell, the schematic letter has changed its significant range of variation in the step from premisses to conclusion, and that is why the reasoning results in fallacy.[13]

It is this sort of complication which justifies the procedure usually adopted in expounding the orthodox hierarchy of types, for it is usual to employ a different style of notation which has no device to parallel my repeating the same letter '$\mathfrak{Q}$' in formulae of different types. The point of my notation is of course to stress that in some cases propositions of different types may be seen as containing the same quantifier, and yet as we have seen this applies only where the quantifier is an unrestricted one. I should admit, then, that the notation I have been using is most naturally employed only when we intend to concentrate our attention on the pure or unrestricted quantifiers, and this I shall shortly do. In fact I shall now introduce a new notation in which the letters 'Q' and 'q' (and their alphabetic variants) will be used to replace the original '$\mathfrak{Q}$' and '$\mathfrak{q}$', and these new letters will be used to range only over the pure quantifiers which are significantly applicable to every monadic propositional form. In the next section I shall give a more or less formal presentation of the logic of these pure quantifiers, while the original '$\mathfrak{Q}$'-notation will lapse, thus bypassing the complications just noticed. But for the present I retain both notations side by side in order to exhibit

[13] Some further remarks on the restrictions involved in a general logic of quantifiers will be found in the appendix to this chapter.

the relation between them more clearly, for indeed the notation for pure quantifiers is very simply introduced as a special case of the notation for quantifiers in general.

It is easy enough to explain any formula that results from a formula of the orthodox hierarchy by substituting 'Q' for '$\mathfrak{Q}$' (and hence by substituting 'q' for '$\mathfrak{q}$'). For example, whereas we said before that a proposition was of the form '$(\mathfrak{Q}x)(Fx)$' if and only if it contained a predicate somewhere, we must now say that if it is to be more particularly of the form '$(Qx)(Fx)$' then in addition the predicate must be contained in a 'type-neutral' way, i.e. in a way in which items of other types may also be contained. More precisely, a proposition is of the form '$(Qx)(Fx)$' if and only if (i) it is of the form '$(\mathfrak{Q}x)(Fx)$', and (ii) there are propositions of the related forms '$(\mathfrak{Q}\mathfrak{f})(\mathfrak{Q}_1 x)(\mathfrak{f}x)$', '$(\mathfrak{Q}\mathfrak{q})(\mathfrak{Q}_2 \mathfrak{f})(\mathfrak{q}x)(\mathfrak{f}x)$',..., and generally of any form '$(\mathfrak{Q}\alpha)(—\alpha—)$' whatever style of variable is put for 'α'. Or, to put the same condition in other words, the proposition must be of some form '(—F—)' written with a predicate-letter somewhere, and in addition the formula '(—F—)' must remain significant upon substitution of any other schema in place of all occurrences of the relevant elementary predicate-schema 'F...', provided that we may also substitute for any *bound variable* already completing the schema 'F...' a new bound variable appropriate to the newly introduced schema, if this is necessary to preserve significance. In a similar way we may modify the original account to explain such related forms as '$(Q_1 x)(Fx)$', '$(Q_2 x)(Fx)$', '$(Q_3 x)(Fx)$',..., and again the related forms '$(Qx)(Fx)$', ('$Qx)(Gx)$', '$(Qx)(Hx)$',...; and a precisely analogous modification is to be made in the explanation of formulae of higher types. There is, then, no difficulty in the introduction of formulae of the orthodox hierarchy in which the quantifier-letters are taken to range only over pure quantifiers. But if we are to be able to speak of quantifiers—and in particular of pure quantifiers—quite generally, no matter what the type of the propositional form they are applied to, then clearly we shall need a more general notation than that of the orthodox hierarchy. So let us start again, still for the moment concerning ourselves with the *general* concept of a quantifier.

The simplest sort of quantifier, and the only sort we have occasion to consider, is a monadic quantifier which yields a proposition when applied to a monadic propositional form.

A proposition containing such a quantifier (initially) can thus be regarded as constructed from two main components, the quantifier and the propositional form, but in spite of this it is not entirely satisfactory to represent the form of that proposition by a formula containing just two schematic letters in juxtaposition (as for instance '$\mathfrak{Q}$' and 'Φ' might be juxtaposed to form the formula '$\mathfrak{Q}(\Phi)$'). For the monadic form to which the quantifier is applied may well have an internal structure that is relevant, and when this is displayed in the symbolism serious ambiguities can arise unless it is explicitly indicated just which of the embedded propositional forms a given quantifier is applied to. (In my example of chapter 3, section 1, we must be able to distinguish between applying the existential quantifier to the monadic forms usually written as '$(\forall x)$(a loves x)' and '$(\forall x)$(x loves b)'; we cannot represent both operations by writing '$(\exists x)(\forall x)$(x loves x)', nor therefore can we represent them by writing '$\exists(\forall$(a loves b))'.) To overcome this difficulty I propose to use repeated variables in the usual way, and so I adopt as the elementary quantifier-formula, representing the form of a proposition constructed by applying a quantifier to a propositional form, the formula '$(\mathfrak{Q}\alpha)(\Phi\alpha)$'—or '$(\mathfrak{Q}_1\beta)(\Psi\beta)$' etc.—where '$\Phi$' and '$\alpha$' play the same syntactic roles as the familiar 'F' and 'x'. The schematic letters 'Φ', 'Ψ',... (with which I associate the variables 'ϕ' 'ψ',...) are to be thought of as themselves representing the whole of a propositional form, and the letters 'α', 'β',... are to be regarded as a sort of dummy variable, differing from ordinary variables in that they do not form part of the specification of the propositional form, but are used *only* to indicate the bondage links. But I shall comment on this departure more fully later.

Now we should consider whether we can give a more formal account of these elementary quantifier-formulae and of the more complicated formulae that we shall wish to construct from them. Just as we first attempted to explain formulae involving subject-quantifiers by relying on all the formulae explained up till then, roughly the formulae of ordinary predicate calculus, so now we might first try to explain our new formulae involving the general notion of a quantifier by relying on all the formulae we have introduced so far, namely the formulae of the orthodox hierarchy of types. For we might say, very simply, that a proposition is of the form '$(\mathfrak{Q}\alpha)(\Phi\alpha)$' if and only if it is also of a form which can be represented by any of the characteristic formulae of the orthodox hierarchy, and more generally for any formula containing occurrences of the new letters 'Φ', 'Ψ',... and 'α', 'β',... we might say that a proposition is of that form if and only if it is also of any form of the orthodox hierarchy which can be obtained from that one by substituting orthodox variables of the same type for 'α', 'β',... and any orthodox schemata of the same type for 'Φ', 'Ψ',....

So understood, the new notation is simply a means for speaking generally of all orthodox types at once. Notoriously Russell held that this could not meaningfully be done, but there seems no reason to agree with him on this point, for in so far as he has any argument here it rests upon the vicious-circle principle which I have already criticized and rejected (in chapter 2, section 1), and besides it is a restriction which Russell is quite unable to observe himself anyway: not only must he speak generally of all types when stating and explaining the theory of types in the first place, but also in his formal development he quite explicitly allows himself the device of 'systematic ambiguity' which just *is* a device for speaking of all types at once. In fact it is his standard practice to announce his theorems and definitions by using letters and variables that strictly range only over one type, but then to add that the theorems and definitions apply also in any other type, provided that the same *intervals* of type are preserved. Thus whereas I have introduced distinct formulae for the characteristic formulae of each type, in *Principia Mathematica* it is very often the case that the formula corresponding to my 'Fa' must be understood as standing indifferently for *any* of my characteristic formulae: 'a' may be taken to be a schematic letter of any type so long as 'F' is taken as a schematic letter of the next higher type. In other more rigourous developments of the theory, such as that due to Church,[14] types are often indicated by adding subscripts to the variables and schematic letters, and what Russell achieves by 'systematic ambiguity' will be achieved rather by the use of schematic subscripts, but whatever device is adopted it must surely be admitted that the procedure is both necessary and legitimate. It simply enables us to sum up an

[14] A. Church, 'A formulation of the simple theory of types', *Journal of Symbolic Logic*, 1940.

infinite set of related theorems, such as

$$\vdash (\forall x)(Fx) \equiv \sim(\exists x)(\sim Fx)$$
$$\vdash (\forall \mathfrak{f})(\mathfrak{Q}x)(\mathfrak{f}x) \equiv \sim(\exists \mathfrak{f})(\sim(\mathfrak{Q}x)(\mathfrak{f}x))$$
$$\vdash (\forall \mathfrak{q})(\mathfrak{Q}_1 \mathfrak{f})(\mathfrak{q}x)(\mathfrak{f}x) \equiv \sim(\exists \mathfrak{q})(\sim(\mathfrak{Q}_1 \mathfrak{f})(\mathfrak{q}x)(\mathfrak{f}x))$$

etc.

in the one formula

$$\vdash (\forall \alpha)(\Phi\alpha) \equiv \sim(\exists \alpha)(\sim\Phi\alpha).$$

For with the present explanation this last formula has exactly the force of the conjunction of all the others: it means simply that the result of applying the universal quantifier to *any* monadic propositional form of the orthodox hierarchy is equivalent to the negation of the result of applying the existential quantifier to the negation of that form.

It is at once clear that, with the present explanation of the new notation, the restriction in this statement to forms *of the orthodox hierarchy* is essential; for not all monadic propositional forms are of the orthodox hierarchy. In fact we have just made use of a propositional form that is not of this hierarchy, namely the form '$(\forall \alpha)(\Phi\alpha) \equiv \sim(\exists \alpha)(\sim \Phi\alpha)$', and we have said of it that any proposition of that form is true. Accordingly we could express this point by using the universal quantifier explicitly,[15] as in

$$(\forall \phi)((\forall \alpha)(\phi\alpha) \equiv \sim(\exists \alpha)(\sim\phi\alpha))$$

and this proposition too is of course equivalent to the proposition

$$\sim(\exists \phi)(\sim((\forall \alpha)(\phi\alpha) \equiv \sim(\exists \alpha)(\sim\phi\alpha)))$$

but the fact that it is so equivalent is not a fact that falls under our original generalization if that concerned only the forms of the orthodox hierarchy. Now it is in fact quite possible to present a formalization of arithmetic without progressing beyond this rather limited interpretation of the new letters 'Φ', 'Ψ',...,[16] but

[15] If the schematic letter 'Φ' is taken as having the same role as a schematic type-subscript, then the quantifier '$(\forall \phi)(\text{—}\ \phi\ \text{—})$' must be regarded as introducing a universal quantification over those subscripts. (And, as I argued in chapter 3, section 1, if it is legitimate to indicate generality by the use of schematic letters of any sort, then it is equally legitimate to introduce quantified variables in place of those schematic letters.)

[16] The axiom of infinity, which was the missing postulate in chapter 1 (p. 15), is easily established with this interpretation of 'Φ' in the form '$\vdash (\exists \phi)(\exists n\alpha)(\phi\alpha)$'. The argument is essentially given by Russell in his *Introduction to Mathematical Philosophy*, chapter 13.

I do not think it would be entirely satisfactory. Not only would we be prevented from making proper use of Frege's theorem, but also we should fall short of a proper understanding of the notion of a number. For numbers, as I urged in chapter 1, are to be understood by reference to 'pure' quantifiers which can be applied unchanged to any propositional form without exception, and that is a conception we have not yet reached.

The point may perhaps be further illustrated by citing an inference of the sort that we shall actually require in the deduction of Frege's theorem. We may say that the propositions of the form '$(\mathfrak{Q}\alpha)(\Phi\alpha)$' that we have mentioned so far, namely those of the orthodox hierarchy, are *first-order* propositions of that form, so that amongst first-order propositions of the form '$(\exists\alpha)(\Phi\alpha)$', for instance, we shall have all propositions of any of the forms

$$(\exists x)(Fx)$$

$$(\exists \mathfrak{f})(\mathfrak{f}a)$$

$$(\exists \mathfrak{q})(\mathfrak{q}x)(Fx)$$

etc.

But in addition we may note that in view of the trivial theorem

$$(\forall\phi)((\exists\alpha)(\phi\alpha) \equiv (\exists\alpha)(\phi\alpha))$$

we are certainly entitled to assert that there is some true proposition of the form

$$(\forall\phi)((\mathfrak{Q}\alpha)(\phi\alpha) \equiv (\exists\alpha)(\phi\alpha)$$

or in other words we are certainly entitled to assert that

$$(\exists\mathfrak{q})(\forall\phi)((\mathfrak{q}\alpha)(\phi\alpha) \equiv (\exists\alpha)(\phi\alpha)).$$

This much we could do even if '$\Phi\alpha$' were restricted to forms of the orthodox hierarchy, but what we could *not* do in that case is recognize the proposition just displayed as *itself* a proposition of the form '$(\exists\alpha)(\Phi\alpha)$', for it is evidently not a first-order proposition of that form. And it is just this sort of recognition that we shall require for the deduction of Frege's theorem.[17]

It does not seem to me difficult to understand the symbol '$\Phi\alpha$'

[17] That is, we shall need to substitute schemata such as

'$(\forall\phi)((\ldots\ \alpha)(\phi\alpha) = (\exists\alpha)(\phi\alpha))$'

for the elementary schema '$\Phi \ldots$', and to recognize that such substitutions preserve validity. See p. 156.

in such a way that this recognition is entirely correct. For the intention of our notation is that a proposition should be said to be of the form '$(\exists\alpha)(\Phi\alpha)$' whenever it contains the existential quantifier initially, no matter what propositional form that quantifier is applied to, and therefore even if the propositional form is one that involves a quantification over all propositional forms. The point seems no more awkward than the point that a proposition of the form '$(\exists\mathfrak{f})(\mathfrak{f}a)$' is itself of the form 'Fa', or that a proposition of the form '$(\exists\mathfrak{q})(\mathfrak{q}x)(Fx)$' is itself of the form '$(\mathfrak{Q}x)(Fx)$', but similarly its consequence is that we have to take our '$\Phi\alpha$' notation as primitive from the formal point of view. The first-order propositions of the form '$(\mathfrak{Q}\alpha)(\Phi\alpha)$' have indeed been introduced already, and so form a basis for the notion which is already at our command. The extensions beyond this basis have been indicated by examples in this section, and described in a general way, so that the projection we are required to make is—it seems to me—altogether natural and easy. And, finally, a general criterion can be laid down which is logically adequate, and which is helpful despite its circularity.

I shall frame this criterion in such a way that it covers the employment of dummy *schematic letters* 'A', 'B',... to answer to the dummy variables 'α', 'β',..., since this will be of some advantage later in the handling of formal proofs. Let us say, then, that a letter from the alphabet 'A', 'B',... is an α-*letter*, a letter from the alphabet 'α', 'β',... is an α-*variable*, and a letter from the alphabet 'Φ', 'Ψ',... a Φ-*letter*. Then where we have any formula containing Φ-letters and (possibly) α-letters and α-variables, we may explain that a proposition is of the form thereby represented if and only if it is also of some form which can be obtained from that one by substituting genuine schematic letters (and not α-letters) uniformly for each α-letter, by substituting genuine variables (and not α-variables) uniformly for each α-variable, and finally by substituting for each Φ-letter a schema (not containing a Φ-letter) which becomes an expression for a propositional form upon completion with a schematic letter. The schema must of course be one that is capable of being significantly completed by the actual schematic letter or variable already substituted for the α-letter and α-variable, and the restriction to be observed is that all α-letters and α-variables that occur in immediate conjunction with the same Φ-letter somewhere in our formula (as in

'ΦA' or 'Φα') should be replaced by genuine schematic letters and variables from the same type—i.e. from the same, or associated, alphabets. (The circularity in the explanation is of course that Φ-*variables* are permitted to occur in the schemata to be substituted for Φ-*letters*, and indeed that Φ-letters and Φ-variables are permitted as substitutions for α-letters and α-variables.)

The restriction upon permitted substitutions for α-letters and α-variables is needed because, as I have said, the α-expressions are mere dummies inserted only to indicate the bondage links. Consequently the task of showing repetition of the same propositional form is performed wholly by repetition of the same Φ-letter, and so we cannot legitimately substitute for one occurrence of such a letter a form written with one schematic letter, and for a second occurrence of the same Φ-letter a different form differing just in being written with a (significantly) different schematic letter. The restriction may appear to be otiose, for at least among the familiar formulae of the orthodox hierarchy of types there are no two which differ *only* in that one contains a schematic letter of one type where the other contains a schematic letter of a different type. But since we are no longer limiting our consideration to formulae of the orthodox hierarchy this fact does not entitle us to leave the restriction tacit, and besides I shall in the next section show how we may indeed introduce formulae which do differ only in this way. So the restriction must be stated explicitly, and as I have already explained it arises because when we analyse a proposition as constructed from a quantifier applied to a propositional form we are analysing it as constructed from two components and not three; hence the 'α' which appears in our elementary formula '(𝔔α)(Φα)' does not represent any propositional component, but only a linkage.

My object in these last pages has been to explain the use of Φ-letters as schematic letters for any monadic propositional form, or more precisely to introduce the idea of a proposition's *containing a monadic propositional form*. I have admitted that this idea is not fully explicable on the basis of the forms of the orthodox hierarchy that we began with, and I regard it as something to be grasped with the help of the copious illustrations in that hierarchy, together with the further examples given in this section and the general (but somewhat metaphorical) discussion of propositional forms that I gave earlier in chapter 2, section 3. But once the notion of a monadic propositional form—or, more briefly, of a Φ-component—is understood, there is of course no further difficulty over the general concept of a quantifier; quantifiers in general are simply to be understood as remainders to Φ-components in just the same way as subject-quantifiers were introduced as remainders to predicates. In fact a proposition is of the form '$(\mathfrak{Q}\alpha)(\Phi\alpha)$' if and only if it is also of any form '$(—\Phi—)$' written with a Φ-letter somewhere (and no quantifier-letter); and the explanation of more complex quantifier forms exactly parallels the explanation already given for subject-quantifiers. In an exactly similar way we may introduce such formulae as '$(\mathfrak{Q}_1\phi)(\mathfrak{Q}_2\alpha)(\phi\alpha)$' to parallel '$(\mathfrak{Q}_1\mathfrak{f})(\mathfrak{Q}_2 x)(\mathfrak{f}x)$', and so on through the whole hierarchy of types. And finally we may as before exchange these formulae for the formulae '$(Q\alpha)(\Phi\alpha)$', '$(Q_1\phi)(Q_2\alpha)(\phi\alpha)$', and so on, where our general quantifier-letters have been replaced by the more narrowly defined pure quantifier letters representing only the quantifiers which are applicable unchanged to all monadic propositional forms. But this change of notation from 'F' to 'Φ' is not *merely* a change of notation, for the new notation is more general in its interpretation than the old, and this is reflected in the more powerful logic which is appropriate to it.

4. The Logic of Pure Quantifiers

In this section I aim to set out, in a fairly formal way, the logic of pure quantifiers that will be used in the following chapters. Except for the purposes of occasional illustration, the general quantifier-letters '$\mathfrak{Q}$', '$\mathfrak{Q}_1$', '$\mathfrak{Q}_2$',... will now lapse, and I shall make no attempt to complete any general logic of quantifiers, since this involves a number of important but rather unusual restrictions, as we saw on pp. 109–10.[18] Indeed the general concept of a quantifier was introduced largely as a step towards the concept of a pure quantifier, in fact as providing a vocabulary with which to introduce the new pure quantifier-letters 'Q', 'Q_1', 'Q_2',..., and explain the use of formulae involving these letters and their associated variables. We may now make use of this explanation

[18] Some further remarks on a general logic of quantifiers will be found in the appendix to this chapter.

to see what logical rules and axioms are appropriate to such formulae, and it is the task of the present section to formulate these rules and axioms explicitly. I proceed in three stages, first introducing what we may call the *basic* logic of pure quantifiers, which is simply isomorphic to the logic of the orthodox hierarchy of types when that is confined to pure quantifiers, next introducing a *first extension* which is made possible by the greater generality of our Φ-symbols, and which makes this logic more powerful than that of the orthodox hierarchy, and finally adding a *second extension*, which is not I think of such great theoretical importance. But I should perhaps begin with a more formal characterization of the formulae we are concerned with, i.e. with the appropriate rules of well-formation.

First, then, besides the usual logical symbols the basic system contains α-symbols, Φ-symbols, and Q-symbols as follows:

α-letters: A, B, Γ... α-variables: α, β, γ...
Φ-letters: Φ, Ψ, X... Φ-variables: φ, ψ, χ...
Q-letters: Q, Q_1, Q_2... Q-variables: q, q_1, q_2...

And it has the following formation rules:

(i) A Φ-letter immediately followed by an α-letter is a well-formed formula

(ii) If $\mathscr{A}$ and $\mathscr{B}$ are any well-formed formulae, then $\ulcorner{\sim}\mathscr{A}\urcorner$, $\ulcorner(\mathscr{A} \& \mathscr{B})\urcorner$, $\ulcorner(\mathscr{A} \vee \mathscr{B})\urcorner$, $\ulcorner(\mathscr{A} \supset \mathscr{B})\urcorner$, and $\ulcorner(\mathscr{A} \equiv \mathscr{B})\urcorner$ are also well-formed formulae

(iii) If $\mathscr{A}$ is any well-formed formula containing a schematic letter, if $\mathscr{B}$ results from $\mathscr{A}$ by replacing some occurrences of that schematic letter by its associated variable, and if $\mathscr{C}$ is the variable in question, then the result of writing either '∀' or any Q-letter in the gap in $\ulcorner(\ldots\mathscr{C})(\mathscr{B})\urcorner$ is a well-formed formula, provided that $\mathscr{B}$ is not already a well-formed formula.[19]

The well-formed formulae of the basic system are just those that can be shown to be well-formed by applying these three rules.[20]

[19] This last proviso rules out 'vacuous' occurrences of quantifiers. (For the use of the quasi-quotation marks '⌜...⌝' see Quine, *Mathematical Logic*, chapter 1, section 6.)

[20] I continue to omit all treatment of relations in this exposition, for I do not think the extra complexity involved would assist in understanding what is important for our purposes. At a later stage I shall in fact be making some small use of dyadic relations corresponding to the monadic 'Φ' (see pp. 149 sqq.), but there will be no difficulty in comprehending this use.

As axioms and rules of inference the system has first some standard basis for propositional calculus and next the axioms and rules given in chapter 3, section 2, as a basis for quantification theory. The last of these rules, rule (iv), needed to be supplemented by special rules appropriate to the treatment of various special propositional components, and these we should now consider. Rule (iv)*a*, which merely allows for alphabetic change of schematic letter or bound variable, is of course to be adopted for all our schematic letters and variables (though it is only in the case of α-symbols that it needs to be stated explicitly). Rule (iv)*b* is concerned with propositional components which are introduced as *remainders* to other components, and this matter requires some discussion.

To begin with, we should notice the Φ-components have effectively been introduced as remainders to *all* propositional components, in the sense that for *any* genuine schematic letter in place of 'A' it will be true that any proposition of any form '(—A—)' is also of the related form 'ΦA'—because evidently such a proposition will be of a form that can be obtained from 'ΦA' by substituting the relevant schematic letter for 'A' and some schema for 'Φ ...'. This entitles us to regard Φ-components as remainders to our *dummy* α-components, for however odd this may appear the essential mark of a remainder for the purposes of rule (iv)*b* is no more than that every proposition of any form '(—A—)' is also of the related form 'ΦA', and this is indeed the case. We may therefore take over the account of Φ-schemata with respect to α-components that was given previously on pp. 70–1, and —remembering the provisos there noted in the account of substitution—we may summarize this rule as the rule that substitution of these Φ-schemata for Φ-letters preserves validity.

The rule of substitution for Q-letters is somewhat more complicated, for we must first spend a little time introducing the notion of a pure Q-schema. This in turn seems most easily done by beginning with the notion of a 𝔔-schema in general, and here we naturally turn for illustration to the 𝔔-schemata that occur in the formulae of the orthodox hierarchy of types.

Now *subject*-quantifiers were introduced as remainders to predicates, in the sense that any proposition of any form '(—F—)' was also of the related form '(𝔔x)(Fx)'. To formulate a rule of substitution for subject-quantifiers, then, we would introduce the idea of a subject-quantifier-schema as what results from any formula of the orthodox hierarchy which contains a predicate-letter by deleting some occurrences of that predicate-letter so as to leave a gap. A subject-quantifier-schema which results in this way from the elementary formula '$(\mathfrak{Q}x)(Fx)$' or its alphabetic variants, e.g. the schema '$(\mathfrak{Q}x)(\text{—}x\text{—})$', we may call an elementary subject-quantifier-schema, and further examples of subject-quantifier-schemata are for instance:

$$(a)\text{:}\quad (\text{—}a\text{—}),\quad (\exists x)(Fx \,\&\, \text{—}x\text{—})$$

$$(b)\text{:}\quad (\forall x)(\text{—}x\text{—}),\quad (\forall x)(\exists y)(Fx \supset \text{—}y\text{—})$$

$$(c)\text{:}\quad (\exists x)(\text{—}x\text{—} \,\&\, (\mathfrak{Q}y)(\text{—}y\text{—} \,\&\, \sim(\forall \mathfrak{f})(\mathfrak{f}x \equiv \mathfrak{f}y))).$$

The relevant rule of substitution (which we would have to spell out in more detail) would then be that substitution of any subject-quantifier-schema for an elementary subject-quantifier-schema preserves validity.

In precisely the same way we would then define a *predicate-quantifier-schema* as what results from any formula of the orthodox hierarchy which contains an elementary subject-quantifier schema by deleting that subject-quantifier-schema so as to leave a gap. The elementary predicate-quantifier schemata would then be the alphabetic variants of '$(\mathfrak{Q}\mathfrak{f})(\text{—}\mathfrak{f}\text{—})$', and other predicate-quantifier-schemata corresponding to the schemata above are:

$$(a)\text{:}\quad (\text{—}F\text{—}),\quad (\exists \mathfrak{f})(\mathfrak{f}a \,\&\, \text{—}\mathfrak{f}\text{—}),\quad (\exists \mathfrak{f})((\mathfrak{Q}x)(\mathfrak{f}x) \,\&\, \text{—}\mathfrak{f}\text{—})$$

$$(b)\text{:}\quad (\forall \mathfrak{f})(\text{—}\mathfrak{f}\text{—}),\quad (\forall \mathfrak{f})(\exists \mathfrak{g})(\mathfrak{f}a \supset \text{—}\mathfrak{g}\text{—})$$

$$(c)\text{:}\quad (\exists \mathfrak{f})(\text{—}\mathfrak{f}\text{—} \,\&\, (\mathfrak{Q}\mathfrak{g})(\text{—}\mathfrak{g}\text{—} \,\&\, \sim(\forall \mathfrak{q})((\mathfrak{q}x)(\mathfrak{f}x) \equiv (\mathfrak{q}x)(\mathfrak{g}x)))).$$

If we were proceeding in the same way through the whole hierarchy of types we would then introduce subject-quantifier-quantifier-schemata as what remained of a formula upon deletion of an elementary predicate-quantifier-schema, and similarly predicate-quantifier-quantifier-schemata, and so on up. But for our purposes it is more convenient to encompass all the remaining $\mathfrak{Q}$-schemata in one blow. For brevity we may refer to the previous sets of schemata as $\mathfrak{Q}x$-schemata and $\mathfrak{Q}\mathfrak{f}$-schemata respectively, and what remains is to introduce $\mathfrak{Q}\mathfrak{q}$-schemata which may be substituted for a general quantifier-letter used to bind a quantifier-variable. This may be done with the help of the idea of a *Φ-schema with respect to* $\mathfrak{Q}$*-components*, which may be defined as what results from any formula containing a $\mathfrak{Q}$-letter by deleting some occurrences of that $\mathfrak{Q}$-letter so as to leave a gap. A $\mathfrak{Q}\mathfrak{q}$-schema is then quite simply defined as what remains of any formula containing a Φ-schema with respect to $\mathfrak{Q}$-components if we delete some occurrences of that Φ-schema so as to leave a gap;[21] the elementary $\mathfrak{Q}\mathfrak{q}$-schemata will then be the alphabetic variants of '$(\mathfrak{Q}\mathfrak{q})(\text{—}\,\mathfrak{q}\,\text{—})$', and other examples to parallel those given before are

$$(a):\quad (\text{—}\,\mathfrak{Q}\,\text{—}),\quad (\exists\mathfrak{q})((\mathfrak{Q}\mathfrak{f})(\mathfrak{q}x)(\mathfrak{f}x)\;\&\;\text{—}\,\mathfrak{q}\,\text{—}),$$
$$(\exists\mathfrak{q})((\mathfrak{q}\mathfrak{f})(\mathfrak{Q}x)(\mathfrak{f}x)\;\&\;\text{—}\,\mathfrak{q}\,\text{—})$$
$$(b):\quad (\forall\mathfrak{q})(\text{—}\,\mathfrak{q}\,\text{—}),\quad (\forall\mathfrak{q}_1)(\exists\mathfrak{q}_2)((\mathfrak{q}_1\,\mathfrak{f})(\mathfrak{f}a) \supset \text{—}\,\mathfrak{q}_2\,\text{—})$$
$$(c):\quad (\exists\mathfrak{q}_1)(\text{—}\,\mathfrak{q}_1\,\text{—}\;\&\;(\mathfrak{Q}\mathfrak{q}_2)(\text{—}\,\mathfrak{q}_2\,\text{—}\;\&\;{\sim}(\forall\mathfrak{q}_3)((\mathfrak{q}_3\,\mathfrak{f})(\mathfrak{q}_1\,x)(\mathfrak{f}x)$$
$$\equiv (\mathfrak{q}_3\,\mathfrak{f})(\mathfrak{q}_2\,x)(\mathfrak{f}x))).$$

Now all the schemata just introduced may be accounted $\mathfrak{Q}$-schemata, but they are not all pure Q-schemata applicable to any monadic form without exception. On the contrary it may seem that *none* of them could be applied to all types of propositional form, since the gaps they contain are already partially filled by a variable of definite type, and this is apt to give the impression that any form the schema could be applied to would have to be a form written with a schematic letter of that same type. However as I was saying earlier on p. 108 when we are thinking from the point of view of the *pure* quantifiers we should not regard the type of the variable bound by a quantifier to be in any way part of the nature of the quantifier itself, for the universal quantifier (for instance) is recognizably the same quantifier whatever the type of the variable that it binds. And this point evidently yields a criterion to determine whether any $\mathfrak{Q}$-schema is more particularly a pure Q-schema, namely that it is so if and only if the variable(s) which appear in that schema somewhere surrounded by a gap could be uniformly replaced by any other variables of any other type while still leaving the

[21] I count a formula as containing occurrences of such a Φ-schema only if at each occurrence of the Φ-schema all of its gaps are filled with the same quantifier-letter or quantifier-variable. When the Φ-schema is deleted, this quantifier-letter or quantifier-variable is of course only written once in the resulting gap.

result a significant $\mathfrak{Q}$-schema. By this criterion it will be seen that the schemata instanced under (*a*) above are not pure Q-schemata, for the first example in each case does not contain a variable at all, while the second does not contain its variable where it could contain any other type of variable. (E.g. if we substitute '$\mathfrak{f}$' for 'x' in '$(\exists x)(Fx \,\&\, \text{—} x \text{—})$' we obtain '$(\exists \mathfrak{f})(F\mathfrak{f} \,\&\, \text{—} \mathfrak{f} \text{—})$', and this is not a $\mathfrak{Q}$-schema since the part '$F\mathfrak{f}$' is ill-formed.) On the other hand the schemata instanced under (*b*) are pure Q-schemata, and this applies to the second example no less than the first since e.g. '$(\forall x)(\exists \mathfrak{f})(Fx \supset \text{—} \mathfrak{f} \text{—})$' is just as satisfactory a $\mathfrak{Q}$-schema as is '$(\forall x)(\exists y)(Fx \supset \text{—} y \text{—})$'. The instance under (*c*) is again not a pure Q-schema by the present criterion, but this is a matter which I shall shortly reconsider.

It will be observed that the $\mathfrak{Q}$-schemata here singled out as pure Q-schemata are precisely those which I relied on earlier (on pp. 104–5) to explain when two propositions should be said to be of such related forms as '$(\mathfrak{Q}x)(Fx)$' and '$(\mathfrak{Q}\mathfrak{f})(\mathfrak{Q}_1 x)(\mathfrak{f}x)$', for what is required is that the first be of some form '(—(F ...)—)' which is obtained by completing a pure Q-schema with the elementary predicate-schema '(F ...)' and that the second be of the related form '$(\text{—}(\mathfrak{Q}_1 x)(\dots x \dots)\text{—})$' which is obtained by taking that same pure Q-schema, changing its variable from 'x' to '$\mathfrak{f}$', and completing it with the elementary subject-quantifier-schema '$(\mathfrak{Q}_1 x)(\dots x \dots)$'. Whether the Q-schema is written with a subject-variable or a predicate-variable (or any other style of variable), I shall accordingly regard it as the same Q-schema in either case, and it is such Q-schemata that will be substitutable for Q-letters in our logic of pure quantifiers.

It is now time to return more directly to that logic. The schemata just discussed were schemata extracted from the formulae of the orthodox hierarchy of types, while our present concern is in fact with a different range of formulae. But the discussion is entirely relevant, for the principles determining what counts as a pure Q-schema are precisely the same whichever range of formulae we start off with. In fact we have only to rewrite that discussion by putting α-letters and α-variables in place of its subject-letters and subject-variables, Φ-letters and Φ-variables in place of its predicate-letters and predicate-variables, and finally Q-letters and Q-variables in place of its general quantifier-letters and quantifier-variables, and then we shall have a perfectly good account of the pure Q-schemata that we shall need in this basic part of the system.[22] So we may now press on to formulate the required rule of substitution for Q-letters.

But first it will be convenient to introduce a little more technical vocabulary. A few pages ago I defined a Φ-schema with respect to Q-components, and we are already familiar with the notion of a Φ-schema with respect to α-components. To complete the account of Φ-schemata I now add a definition of a Φ-schema *with respect to Φ-components*, for this is simply what results from any formula containing a Φ-letter when some occurrences of that Φ-letter are deleted so as to leave a gap. (A Φ-schema with respect to Φ-components is therefore the very same thing as a 𝔔α-schema, i.e. the analogue in the present system to a 𝔔x-schema in the orthodox hierarchy.) Further, I introduce the notion of a *partially* formed formula as including both well-formed formulae and what results from any well-formed formula by exchanging some of its schematic letters for their associated variables; and similarly I introduce a *partial* Φ-schema (with respect to any sort of components) as including both Φ-schemata (with respect to those components) and what results from such a Φ-schema by exchanging some of its schematic letters for their associated variables. Finally, we may define the *scope* of a given occurrence of a Q-letter as the shortest partially formed formula in which it occurs, and we may regard this scope as being composed of an elementary Q-schema, consisting of the Q-letter itself and the variable which it binds, and then a partial Φ-schema which becomes a partially formed formula when the variable in question is restored to its various gaps. So I shall say that the elementary Q-schema is *completed by* this partial Φ-schema at this occurrence.

Thus in any formula containing a Q-letter each occurrence of that Q-letter will have a scope, and in each case the scope will consist of the same elementary Q-schema (written with either an α-variable or a Φ-variable or a Q-variable), but completed by different partial Φ-schemata (some perhaps being Φ-schemata with respect to α-components, some with respect to Φ-components, and some with respect to Q-components). Now suppose

[22] In the extension to follow more pure Q-schemata will be introduced.

that in such a formula we replace each occurrence of the elementary Q-schema with an occurrence of a new pure Q-schema, the same new Q-schema for each occurrence, but in such a way that this newly introduced Q-schema is each time written with the same style of variable, and completed with the same partial Φ-schema, as was the corresponding occurrence of the original elementary Q-schema; then provided that the quantifiers in the newly introduced Q-schema do not anywhere capture any of the variables originally contained in our formula (so disturbing its bondage links), we may say that the new formula results from the old by substituting a Q-schema for a Q-letter, and the rule is that such substitutions preserve validity.

With these two specifications of rule (iv)*b*, the one permitting substitution for Q-letters of any pure Q-schema, and the other permitting substitution for Φ-letters of any Φ-schema with respect to α-components, the presentation of the *basic* logic of pure quantifiers is completed.

It will be seen that the logic as developed so far is strictly isomorphic to the logic of the orthodox hierarchy of types when that is confined to pure quantifiers. Indeed it could be said that so far we appear to have introduced a merely notational change, exchanging 'x' for 'α' and 'F' for 'Φ', but there is no reason to suppose that our different notation brings any extra inferential power. However, there is some extra power yet to be added, just because the intended interpretation of our Φ-letters is a great deal more general than the intended interpretation of predicate-letters, so that a single theorem of this logic, such as

$$\vdash (\forall\alpha)(\Phi\alpha \vee \sim\Phi\alpha)$$

states at one blow all that can be expressed even by an infinity of theorems of the orthodox hierarchy (where we must use one theorem for each type) and yet more besides. To bring this extra generality into the open we might therefore go on to introduce some rules linking the present system to the orthodox one, in particular by making explicit the fact that the schematic letter 'Φ', which stands in for *any* monadic propositional form, does inter alia stand in for any monadic form written with a schematic subject-letter, and for any monadic form written with a schematic predicate-letter, and so on.

A convenient way of doing this would be to introduce the new formula 'Φa'—or, in quantified contexts, 'Φx', 'ϕx', and 'ϕa'—specifically to represent a monadic form written with a schematic subject-letter, and in these formulae we should regard the symbols 'a' and 'x' as having the dual role both of specifying the letter 'Φ' in this way and of supplanting the α-letters and α-variables in indicating the bondage links. Clearly 'Φa' and 'Φx' are merely notational variants on 'Fa' and 'Fx', and they could perfectly well be introduced as such, but they have the advantage of allowing a particularly simple formulation of a rule connecting the present system with that of the orthodox hierarchy. This rule, though it is really a further rule of substitution for Φ-letters, we may conveniently call a rule of substitution for α-symbols; for evidently what holds for all monadic forms holds also for all monadic forms written with a subject-letter, and so we shall preserve validity if we take any formula containing a Φ-letter occurring in conjunction with α-letters and α-variables and substitute for it the corresponding formula in which those α-letters and α-variables have been supplanted by subject-letters and subject-variables. In the same way we would go on to introduce 'ΦF' to represent any monadic form written with a schematic predicate-letter, and so as a variant on '$(\mathfrak{Q}x)(Fx)$', and the same substitution-rule would equally permit us to supplant our α-letters and α-variables by predicate-letters and predicate-variables. Pressing on in the same way again we could then introduce an analogous formula, say '$\Phi\underset{x}{\mathfrak{Q}}$', to represent any monadic form written with a schematic subject-quantifier-letter, and therefore as a variant on '$(\mathfrak{Q}_1 \mathfrak{f})(\mathfrak{Q}x)(\mathfrak{f}x)$'; and so on throughout the hierarchy. In this way we would show how the present symbolism transcends that of the orthodox hierarchy, and how the logic we are now concerned with effectively includes the logic of that hierarchy. But I shall not pause to formulate these cases of the rule of substitution for α-symbols more precisely, for although the basic logic of quantifiers can indeed be properly extended in this way we shall have no occasion to make use of such an extension in what follows. Besides, it will be very easy to see how these rules should be precisely formulated from the very similar extension that I shall now introduce.

For just as we may consider the effect of limiting our schematic letter 'Φ' to propositional forms written with a schematic subject-letter, or predicate-letter, and so on, so we may also consider the effect of limiting 'Φ' to propositional forms written with a schematic Φ-letter or with a schematic Q-letter, and this latter we shall require to do. If we say that in the formulae of the basic system we can recognize α-positions, Φ-positions, and Q-positions as being those positions which can significantly be filled by α-symbols, Φ-symbols, and Q-symbols respectively, then the extension of the basic system which I shall now introduce is one which contains new formulae wherein Φ-symbols, and Q-symbols may occur not only in 'their own' positions but also in α-positions.

Now just as the notation 'ΦF' could be introduced as a variant on the notation '(𝔔x)(Fx)', so equally the notation 'ΦΨ' can be understood as a variant on the notation '(𝔔α)(Ψα)'; but since the general quantifier-letter '𝔔' does not actually occur in the system we are presently considering this point is of no great use to us, and we must in fact introduce the new notation as primitive. It may seem that this notation is bound to lead to Russellian contradictions, for one naturally expects that if 'ΦΨ' is to be admitted as a well-formed formula then 'ΦΦ' must also be admitted, and this will surely lead to the same difficulties as 'FF' would lead to in the orthodox systems of logic. However, this is not so. If one thinks what the new notation means it becomes clear that 'ΦΦ' must remain ill-formed even though 'ΦΨ' is admitted.

A formula such as 'Φa' or 'ΦF' or 'ΦΨ' is as a whole a schematic expression for a monadic propositional form, and while the letters 'a' or 'F' or 'Ψ' in these formulae serve to show that the propositional form in question is one that is written with a schematic subject-letter or predicate-letter or Φ-letter (as the case may be), still it is really the initial letter 'Φ' that stands in for the whole propositional form. That is why I emphasized in the last section (p. 117) that repetition of the same Φ-letter is itself enough to indicate repetition of the same propositional form, and it follows from this that no formula can significantly contain both 'Φa' and 'ΦF'—for evidently if a monadic form is written with a subject-letter as its schematic letter then it is not written with a predicate-letter as its schematic letter. (On the other hand there is no bar to a formula containing both 'Φa' and 'Φb', for the monadic form 'a is white', for instance, is the very same form as the form 'b is white'.) To speak more concisely, let us say that when an occurrence of a Φ-letter (in Φ-position) is one that is immediately followed not by a dummy α-symbol but by some genuine variable or schematic letter, then the Φ-letter is *limited* in that occurrence, for its range of variation is limited to forms written with the type of schematic letter in question. The principle just noticed, then, is that no formula can contain the same Φ-letter limited one way in one occurrence and another way in another occurrence. Again, we may say that an occurrence of a Φ-letter (in Φ-position) which is followed by a dummy α-symbol is *not* a limited occurrence, just because the α-symbols are mere dummies and so permit the Φ-letter to range over all monadic forms without exception. Then similarly we must rule that no formula can contain the same Φ-letter limited in one occurrence and unlimited in another, for even though the two ranges of variation overlap we have seen (pp. 109–10) that it will lead to fallacies if we change a range of variation in mid-formula. Now the relevance of this is that where a Φ-letter occurs *in α-position*, as 'Φ' occurs in 'ΨΦ', it must be regarded as unlimited in that occurrence, for its role is to specify merely that the monadic form represented by the whole formula is one that is written with a Φ-letter as its schematic letter, but no limitation is placed on the range of variation of that schematic Φ-letter. And it follows from this that 'ΦΦ' cannot be a well-formed formula, because the first occurrence of 'Φ' is limited and the second is not.[23]

But though the same Φ-letter cannot occur both in an α-position and in a Φ-position where it is limited, there is nothing against repeating the same Φ-letter both in an α-position and in an unlimited Φ-position, as in '(∀α)(Φα) & ΨΦ'. Clearly what this formula represents is any propositional form which is built up by conjoining the formula '(∀α)(Φα)' with any other formula which contains that same schematic letter 'Φ' somewhere, and repetition of the same letter 'Φ' may still be interpreted as indicating repetition of the same propositional form. In fact a proposition is of the form '(∀α)(Φα) & ΨΦ' if and only if it may be analysed as a conjunction of two propositions as components, the first of which is the proposition that all propositions of a certain monadic form are true, while the second is a proposition that contains that same monadic form somewhere. The same type of explanation can evidently be extended to more complex formulae

[23] Further discussion of the formula 'ΦΦ' will be found in the appendix to this chapter, p. 139.

as is clear from the point made at the outset, that the formula '$\Psi\Phi$' can be understood as a mere notational variant on '($\mathfrak{Q}\alpha$) ($\Phi\alpha$)'.

No extra complication arises when we turn to consider the use of Q-letters in α-positions. We must of course understand the occurrence of 'Φ' in 'ΦQ' to be a *limited* occurrence, but that point has already been catered for, and there is no similar distinction to be drawn for the various occurrences of 'Q'. A pure quantifier-letter will always range over all pure quantifiers, no matter what its context, just because pure quantifiers are applicable unchanged to all monadic forms. And finally, it may be observed that the explanation given on pp. 116–17 of formulae containing α-symbols in α-position carries over without change to the new formulae with Φ-symbols and Q-symbols in α-position, so I think we may now proceed to a formulation of the rules and axioms of this first extension to the basic system.

First, the original formation-rules must clearly be altered, and rule (i) must be extended to

(i)′: A Φ-letter immediately followed by an α-letter or a Q-letter or a different Φ-letter is a well-formed formula

while rule (ii) must be restricted to

(ii)′: If $\mathscr{A}$ and $\mathscr{B}$ are well-formed formulae, then $\ulcorner{\sim}\mathscr{A}\urcorner$, $\ulcorner(\mathscr{A}\ \&\ \mathscr{B})\urcorner$, $\ulcorner(\mathscr{A} \vee \mathscr{B})\urcorner$, $\ulcorner(\mathscr{A} \supset \mathscr{B})\urcorner$, $\ulcorner(\mathscr{A} \equiv \mathscr{B})\urcorner$ are also well-formed formulae, provided that all Φ-letters common to both $\mathscr{A}$ and $\mathscr{B}$ are limited in the same way if they are limited at all—i.e. provided that any Φ-letter which occurs in $\mathscr{A}$ (or $\mathscr{B}$) immediately followed by a Φ-symbol or by a Q-symbol, and which also occurs in $\mathscr{B}$ (or $\mathscr{A}$), occurs in $\mathscr{B}$ (or $\mathscr{A}$) immediately followed by a Φ-symbol again or by a Q-symbol again.

Second, the restriction in the new rule (ii)′ may also give rise to a restriction in the underlying propositional logic, simply to ensure that all theorems are well-formed. If that logic is formulated with Modus Ponens as its only rule of inference then no change need be made, but if for instance the propositional logic takes as primitive the rule of adjunction, that if $\vdash\mathscr{A}$ and $\vdash\mathscr{B}$ then $\vdash\ulcorner\mathscr{A}\ \&\ \mathscr{B}\urcorner$, then we must clearly restrict that rule to cases where $\ulcorner\mathscr{A}\ \&\ \mathscr{B}\urcorner$ is well-formed by the new clause (ii)′. Again the rules for first-order quantification theory may have to be restricted for the same reason, for obviously we must not be permitted to pass from the well-formed '$\vdash (\forall\phi)(\phi A \supset (\phi A \vee \Psi Q))$' to the ill-formed '$\vdash \Psi A \supset (\Psi A \vee \Psi Q)$'. So it seems simplest to regard this restriction as a blanket restriction on all the usual rules of inference. Otherwise the underlying propositional logic, and the quantification theory given by rules (i)–(iv)*a* of chapter 3, section 2, remain unchanged.

But there are changes in the specifications of rule (iv)*b*. The previous rule of substitution for Φ-letters (in Φ-position) concerned only the possibility of substituting for a Φ-letter that was immediately followed by α-symbols any Φ-schema with respect to α-components. This rule of course remains, but we now add in addition that we may substitute for any Φ-letter that is immediately followed by Φ-symbols or by Q-symbols any Φ-schema with respect to Φ-components or Q-components respectively. (Whereas the earlier rule states in effect that Φ-components are remainders to all propositional components, these two extensions state that Φ-components are in particular remainders to Φ-components and to Q-components.) The definitions of these Φ-schemata remain as before, viz as what results from any well-formed formula containing a Φ-letter (or a Q-letter) upon deletion of some occurrences of that letter so as to leave a gap, though the effect is slightly different since we now have an enlarged totality of well-formed formulae to consider.

The rule of substitution for Q-letters (in Q-position) now requires a more careful formulation. It is still true, of course, that any pure Q-schema may be substituted for such a Q-letter, and it is still correct to regard a pure Q-schema as being a $\mathfrak{Q}$-schema in which the type of the relevant variable(s) is irrelevant. But it is no longer sufficient to say that this type is irrelevant so long as a change to variables of another type still yields a significant $\mathfrak{Q}$-schema. For if we consider such a $\mathfrak{Q}$-schema as

$$(\exists\alpha)(\Phi\alpha \,\&\, \text{—}\alpha\text{—})$$

then a change to variables of another type will yield

$$(\exists\psi)(\Phi\psi \,\&\, \text{—}\psi\text{—})$$

or

$$(\exists q)(\Phi q \,\&\, \text{—}q\text{—})$$

and these are significant $\mathfrak{Q}$-schemata in the present extended system. But it will not do to regard them as simply different ways of writing the same pure Q-schema, because the schematic letter 'Φ' which they contain cannot be interpreted in the same way in each of them. Indeed if we attempted to substitute this alleged Q-schema for the schematic letter 'Q' in

$$(Q\alpha)(\Psi\alpha)\ \&\ (Qq)(Xq)$$

we should, by the previous explanation, obtain

$$(\exists\alpha)(\Phi\alpha\ \&\ \Psi\alpha)\ \&\ (\exists q)(\Phi q\ \&\ Xq)$$

which is not well-formed, because it contains the same letter 'Φ' now limited and now unlimited.

To overcome this difficulty we must therefore redefine a pure Q-schema as any $\mathfrak{Q}$-schema which (i) yields further significant $\mathfrak{Q}$-schemata when the type of the variable(s) that occur in it somewhere surrounded by gaps is uniformly altered, and (ii) which could occur together with these further significant $\mathfrak{Q}$-schemata in a well-formed formula. But perhaps it is more perspicuous to point out that according to this definition a $\mathfrak{Q}$-schema whose relevant variables are α-variables will also be a pure Q-schema if and only if the α-variables that occur in its gaps have no 'chain of connection' with a schematic Φ-letter or a schematic α-letter—i.e. if they do not occur in immediate conjunction with a Φ-variable which (elsewhere) occurs in immediate conjunction with an α-variable which (elsewhere) occurs in immediate conjunction with a Φ-variable which (elsewhere) occurs . . . in immediate conjunction with a schematic Φ-letter or a schematic α-letter.

It will be seen that the $\mathfrak{Q}$-schemata of the present system which correspond to the first examples of $\mathfrak{Q}$x-schemata given on p. 121 are still not pure Q-schemata by this ruling, while those corresponding to the second examples still are. But for the sake of future developments it is important to note here that although the third example of a $\mathfrak{Q}$x-schema, viz.

$$(\exists x)(\text{—}x\text{—}\ \&\ (\mathfrak{Q}y)(\text{—}y\text{—}\ \&\sim(\forall \mathfrak{f})(\mathfrak{f}x \equiv \mathfrak{f}y)))$$

was not a case of a pure Q-schema, nevertheless the schema that now corresponds to this, viz.

$$(\exists\alpha)(\text{—}\alpha\text{—}\ \&\ (Q\beta)(\text{—}\beta\text{—}\ \&\sim(\forall\phi)(\phi\alpha \equiv \phi\beta)))$$

is a pure Q-schema. For if we now substitute Φ-variables or Q-variables for the α-variables in this schema then we obtain further schemata which do count as well-formed by the extended rules of well-formation and which can all occur in the same well-formed formula. What made the previous case inadmissible was that '$\Phi\alpha$' and 'Φq' could not both occur in the same formula, but there is no reason why the same formula should not contain both '$(\forall\phi)(\phi\alpha = \phi\beta)$' and '$(\forall\phi)(\phi q_1 \equiv \phi q_2)$', for although the range of the variable 'ϕ' is indeed different in the two clauses it is not governed by the same quantifier in both clauses, and so no awkwardness will result. It may perhaps seem that if the variable does have a different range in the two clauses then it is somewhat misleading to say that the substitution of one clause for another makes no difference to the quantifier thereby expressed, but I think this impression vanishes on closer reflection. For we should recall that sameness of quantifier was originally explained by saying that propositions contained the same quantifier (initially) when they were built up in the same way from certain initial components, and it does seem entirely proper to say that two propositions respectively of the forms

$$(\exists\alpha)(\Phi\alpha \,\&\, (Q\beta)(\Phi\beta \,\&\, \sim(\forall\phi)(\phi\alpha \equiv \phi\beta)))$$
$$(\exists q_1)(\Psi q_1 \,\&\, (Qq_2)(\Psi q_2 \,\&\, \sim(\forall\phi)(\phi q_1 \equiv \phi q_2)))$$

are indeed built up in the same way, the first being built up from some monadic form or other (here represented by 'Φ') and the second being built up in the same way from a monadic form (represented by 'Ψ') which has a quantifier-letter as its schematic letter. Besides, on an intuitive level we may remark that the clause '$(\forall\phi)(\phi q_1 \equiv \phi q_2)$' says in effect that any proposition containing q_1 has the same truth-value as the corresponding proposition containing q_2, and surely this same thing can be said not only of propositions containing quantifiers but also of propositions containing any other type of component.

This completes my account of how the original rules of our basic system are to be modified for the extended system now under consideration, but we have not yet formulated the new rule which is really what is characteristic of this extended system, namely the rule of substitution for α-symbols. (As I have said, the rule is really a further rule of substitution for Φ-letters, but it is convenient to give it a different title). Suppose, then, that we take any formula containing some schematic Φ-letters which are immediately followed by α-symbols and that we select some of those Φ-letters and simultaneously substitute Q-letters or Φ-letters uniformly for all occurrences of the α-letters that have some occurrence immediately following one of the selected Φ-letters, and Q-variables or Φ-variables uniformly for all occurrences of the α-variables that are bound to the same quantifier as some occurrence of an α-variable immediately following one of the selected Φ-letters; then if the resulting formula is well-formed (and if it preserves the same bondage-links), we may say that this new formula results from our original by substitution for α-symbols, and the new rule that we require is of course that such substitutions preserve validity.

The first extension of our basic logic of pure quantifiers is now complete.

The need for a further extension can be seen when we reflect that so far we have introduced rules allowing us to substitute Φ-schemata for Φ-letters, and Q-schemata for Q-letters, *but* only when these letters occur in Φ-position or in Q-position respectively. We have introduced no rules which permit substitutions for letters in α-position except the trivial rule (iv)*a* which concerns alphabetic variants. Indeed so far as our first extension is concerned no further rules of this sort would be of any use, for since the only symbols permitted to occur in α-position are single variables or schematic letters, if we did attempt to substitute the constant '∀' for the schematic letter 'Q' in the context 'ΦQ', then the resulting formula 'Φ∀' would not even be well-formed. But clearly if 'ΦQ' represents any monadic propositional form written with a schematic quantifier-letter, then it will be entirely natural to take 'Φ∀' as representing the proposition of that form which is got by writing '∀' in place of the schematic letter, and accordingly I shall so take it.

Now the system we are concerned with contains only one primitive quantifier constant, viz. '∀', but of course it also contains the means of expressing further such constants, for instance by the prefix '∼(∀ ...)∼', and if '∀' can legitimately follow a Φ-letter then we should clearly allow these other quantifier constants to do so too. In fact the same point arises quite generally with regard to any pure quantifier schema, whether it is the expression for a constant or whether—because it contains schematic letters within itself—it is again a schematic quantifier expression. For instance we might possibly be interested in all quantifiers which were expressed by a prefix of the form '$\sim$(Q ...)$\sim$', and then it would be appropriate to use a Φ-letter followed by this prefix (or something like it) to stand in for any monadic form which had a quantifier-letter as its schematic letter and which contained that quantifier-letter in the manner indicated. So the second extension to the basic system is an extension in which a Φ-letter may be followed not only by a single schematic letter (or variable) but also, in effect, by any schema or constant that is substitutable for such a schematic letter.

However, I shall not present this extension in full. I shall quite ignore the possibility of using Φ-schemata in α-position, because indeed in what follows there will not even be any occasion to use Φ-letters (or variables) in α-position, and this part of the development was introduced mainly for the sake of clarifying the notion of a pure Q-schema. As regards the use of Q-schemata in α-position, this will be confined to so few examples that I think it is not worth while introducing the cumbersome formulae that would strictly be required, and so I propose to adopt an abbreviating technique. (As a matter of fact our present, limited, purposes would be adequately served if we introduced the new notation we require by an explicit definition modelled on Russell's definition of classes, for example, defining

$$\text{'}\Phi\forall\text{' for '}(\exists q)((\forall\psi)((\forall\alpha)(\psi\alpha) \equiv (q\alpha)(\psi\alpha)) \;\&\; \Phi q)\text{'}.$$

But in default of an axiom of extensionality—which I do not provide—it is doubtful whether this definition would preserve the sense really intended, and therefore I prefer to recognize a further extension to the logic at this point.[24]

The abbreviating technique may be introduced with the help of the notion of a Q-*expression* (which includes a Q-letter as a special case). By a Q-*expression* I mean any expression, whether primitive *or defined*, such that (i) when inserted into the gap in (... α)(Φα) it yields either a well-formed formula or an abbreviation for a well-formed formula, and (ii) the well-formed formula which this expression is or abbreviates is obtainable from the formula '(Qα)(Φα)' by substitution for Q-letters. A Q-expression is therefore an expression with the same syntactical role as a

[24] This extension is further discussed in the appendix, p. 138.

Q-letter, and it is always either a constant or a schematic expression for a quantifier. It is these Q-expressions, then, which we shall permit to occur in the positions formerly occupied by Q-letters. Of course it will be a consequence of our adopting this technique that we offend against the principle that a defined sign should be eliminable from any context in which it occurs; for example, although '$\exists$' will be eliminable from any context in which it occurs in Q-position, it will not be eliminable from contexts in which it occurs in α-position. To put this right, we must strictly regard '$\Phi\exists$' as an abbreviation for, say, '$\Phi(\lambda\psi(\exists\alpha)(\psi\alpha))$', and that in turn as an abbreviation for '$\Phi(\lambda\psi(\sim(\forall\alpha)(\sim\psi\alpha)))$', which last we take to be well-formed by the rules of well-formation. The procedure I adopt simply short-cuts these rather cumbersome formulae.

The formal presentation of this second extension to the basic system need not detain us long. First we retain the definition of 'Q-schema' previously given in terms of the formulae of the first extension, and with the help of that definition we define 'Q-expression' as above. Then we add a new clause to the recursive specification of well-formed formulae stating that a Φ-letter followed by a Q-expression is a well-formed formula, we note that the rule of substitution for Φ-letters when followed by Q-symbols now applies also to Φ-letters when followed by Q-expressions, and finally we extend the rule of substitution for Q-letters by allowing that substitution of Q-expressions for Q-letters preserves validity, not only when those Q-letters occur in Q-position but also when they occur in α-position.

This completes my presentation of the logic of pure quantifiers, or at least of as much of that logic as will be used in what follows. As a matter of fact only some of the logical principles I have been presenting will be used hereafter, and perhaps it will be helpful[25] if I close this chapter with a brief summary of what they are.

First, then, the formulae which I shall be employing contain quantifier-constants used to bind α-variables, Φ-variables, and Q-variables, but schematic quantifier-letters are used only to

[25] Helpful at least in assessing the soundness of the deduction. For obviously there is a danger that a logic that is not bound by Russellian type theory will prove inconsistent in unexpected ways, and I have no proof that the logic here presented is consistent.

bind α-variables and Q-variables. Φ-variables are not needed in this role, and as I have mentioned Φ-symbols do not appear in α-position either. Q-symbols do frequently occur in α-position and consequently I make use of the rule permitting substitution of Q-symbols for α-symbols. So far as concerns the rule of substitution for Φ-letters, I invoke this rule to justify substitution of Φ-schemata with respect to α-components (in the manner discussed on pp. 70–2) and of Φ-schemata with respect to Q-components, but in the latter case what is substituted for 'ΦQ' will either contain 'Q' once more in α-position or if 'Q' occurs in Q-position it will be binding an α-variable. (Hence, although I use formulae in which a Q-letter binds a Q-variable, I do not have any occasion to use, when substituting for Φ-letters, the Φ-schemata with respect to Q-components that could be extracted from such formulae.)[26] Finally, the rule of substitution for Q-letters is only used in two cases, which arise in connection with the definitions

$$`(\exists\alpha)(\Phi\alpha)' \quad \text{for} \quad `{\sim}(\forall\alpha)({\sim}\Phi\alpha)'$$

$$`(Q'\alpha)(\Phi\alpha)' \quad \text{for} \quad `(\exists\alpha)(\Phi\alpha \,\&\, (Q\beta)(\Phi\beta \,\&\, {\sim}(\forall\phi)(\phi\alpha \equiv \phi\beta)))'.$$

The expressions '∃' and 'Q′' introduced by these definitions are Q-expressions in the sense defined on p. 134, and I invoke the rule to licence their substitution for schematic Q-letters.

With the expression 'Q′' just defined we are clearly approaching the topic of numerical quantifiers, and the clause '$(\forall\phi)(\phi\alpha \equiv \phi\beta)$' in this definition at once raises the question of the connection between the notion of a numerical quantifier and that of identity. So let us now move on to consider this question.

Appendix. Some Remarks on a General Logic of Quantifiers

The fallacy discussed on pp. 109–10 stems from the assumption that the usual rules of propositional logic will apply unchanged to formulae involving general quantifier-letters, and the reason why this assumption fails is because although the general quantifier-letters are used to stand in for all varieties of quantifiers—both pure quantifiers and quantifiers restricted to this or that type—nevertheless their significant range of variation alters from one context to another: in particular, wherever '$\mathfrak{Q}$'

[26] The significance of this point emerges in the appendix to this chapter, p. 140.

occurs binding subject-variables it must be taken as ranging over no more than the subject-quantifiers, wherever '$\mathfrak{Q}$' occurs binding predicate-variables it must be taken as ranging over no more than the predicate-quantifiers, and wherever '$\mathfrak{Q}$' occurs binding both subject-variables and predicate-variables it must be taken to range only over those quantifiers which are both subject-quantifiers and predicate-quantifiers, namely the pure quantifiers. In order to avoid the fallacies we must therefore place some restriction on the application of propositional logic to these general quantifier formulae, in fact we must restrict the application to formulae in which the same $\mathfrak{Q}$-letter has the same range of variation throughout. But of course if what we are concerned with is a *logic*, that is to say with rules of inference for these formulae, we shall not be in any way interested in the formulae to which propositional logic is not applicable, and so the simplest procedure will be to dismiss them at the outset as ill-formed.[27]

One way of achieving this would be to allow a general $\mathfrak{Q}$-letter to occur in a formula only when in each of its occurrences it binds a variable of the same type, that is (in the case of the orthodox hierarchy) when in each occurrence it binds a subject-variable, or when in each occurrence it binds a predicate-variable, or when in each occurrence it binds a quantifier-variable which in turn binds a subject-variable, and so on. But since we also wish to allow for the existence of pure quantifiers, capable of binding any type of variable, we should then have to add a further set of formulae containing pure Q-letters, and not subject to any such restriction. Thus although, according to my explanations, there are propositions of such a form as '$(\mathfrak{Q}x)(Fx)$ & $(\mathfrak{Q}\mathfrak{f})(\mathfrak{f}a)$', we are as it were not allowed to say so for the purposes of logic, but must content ourselves with the observation that all such propositions are more particularly of the form '$(Qx)(Fx)$ & $(Q\mathfrak{f})(\mathfrak{f}a)$', for this second is a formula that we are permitted to use. The part of the logic which merely involves $\mathfrak{Q}$-letters will then be fairly close to the more ordinary treatments of the orthodox hierarchy, since we shall have removed what was the main distinguishing feature of my treatment, viz. the possibility of repeating the same $\mathfrak{Q}$-letter in typically different contexts. (But there is still no bar to formulae related as '$(\mathfrak{Q}_1 x)(\mathfrak{Q}_2 \mathfrak{f})(\mathfrak{f}x)$' and '$(\mathfrak{Q}_2 \mathfrak{f})(\mathfrak{Q}_1 x)(\mathfrak{f}x)$', and these have no analogue in the usual treatments.) But alongside this part, whose distinctive rules are those quickly sketched on pp. 121–2, we shall also have a part dealing with Q-letters with rules as given in the last section, and we shall in addition have a linking rule which makes 'Q' substitutable for '$\mathfrak{Q}$', since evidently what holds for all quantifiers applicable to certain variables holds also for all pure quantifiers when applied to those variables.

I hope this rather brief description is sufficient to show how one could set out a general logic of quantifiers for handling the formulae of the orthodox hierarchy. When we come to remove the limitation to this hierarchy, then as our first stage we proceed, as before, by simply

[27] It was precisely this method that was employed with Φ-letters on pp. 127–30 when these too were used with different significant ranges of variation. In fact this use of Φ-letters is almost the same thing as the use of general quantifier-letters, as will appear more fully in a moment.

rewriting 'F' and 'x' everywhere as 'Φ' and 'α', but retaining precisely the same rules of inference. In the second stage we have to add the rule which shows the greater generality of our Φ-letters, namely the rule of substitution for α-symbols. But whereas previously this new rule could be introduced only if we first introduced a new range of formulae in which α-symbols were supplanted by other symbols, in the present case the new formulae required can largely be introduced as simple abbreviations of some of the old ones, or in other words they can in most cases be by-passed altogether. For I did remark that the formula '$\Phi\Psi$' could be understood simply as a notational variant on the formula '$(\mathfrak{Q}\alpha)(\Psi\alpha)$', and this latter formula is in the present case already to hand. In fact we could introduce the notation with Φ-symbols in α-position by means of the abbreviation rule: any formula which contains an elementary $\mathfrak{Q}\alpha$-schema (i.e. '$(\mathfrak{Q}\alpha)(\text{—}\alpha\text{—})$' and its alphabetic variants) which is completed at each of its occurrences by a single Φ-symbol may be abbreviated by writing a new Φ-letter uniformly in place of the $\mathfrak{Q}\alpha$-schema and writing immediately after each occurrence of the new Φ-letter the same Φ-symbol as was used to complete the corresponding occurrence of the original $\mathfrak{Q}\alpha$-schema. (The newly introduced Φ-letter must of course be one that does not already occur elsewhere in the original formula, for repetitions of this new Φ-letter serve to abbreviate repetitions of the $\mathfrak{Q}\alpha$-schema, and these originally had nothing in common with any Φ-letters occurring in the formula.) This rule introduces '$\Phi\Psi$' and '$\Phi\psi$' and their alphabetic variants and a precisely similar rule may be given for '$\phi\psi$' and '$\phi\Psi$' and their alphabetic variants, so we now have our formulae with Φ-symbols in α-position introduced by explicit definition.

The rule permitting substitution of Φ-symbols for α-symbols may now be restated, and indeed extended. Previously we permitted such substitution for α-symbols that occurred in conjunction with schematic Φ-letters, on the ground that an α-symbol is a mere dummy and a Φ-letter, which represents any propositional form, does *inter alia* represent those forms that are written with a schematic Φ-letter. But now we can add that the rule also applies to α-variables that are bound by a schematic $\mathfrak{Q}$-letter, on the very same ground that the α-variable is a mere dummy and the $\mathfrak{Q}$-letter, which represents any quantifier, does *inter alia* represent quantifiers which are restricted to Φ-variables; and to say that a quantifier is restricted to Φ-variables is of course just the same as to say that the propositional forms to which it is applied must be ones that are written with a schematic Φ-letter. So the rule of substitution for α-symbols, which was really an extra rule of substitution for Φ-letters, is now extended to form a similar extra rule of substitution for $\mathfrak{Q}$-letters as well.

Before passing on to consider quantifier-symbols in α-position there are two points worth noting at this stage. First, although we had previously to recognize a *second* extension of our basic system whereby we moved from allowing single letters and variables to occur in α-position to a more liberal toleration of any corresponding schemata, this second extension has now become quite superfluous. For just as '$\Phi\Psi$' is introduced as a variant on '$(\mathfrak{Q}\alpha)(\Psi\alpha)$', so now we could—if there were any need for it—

introduce say '$\Phi(\lambda\alpha(\Psi\alpha \vee \sim\Psi\alpha))$' as a variant on '$(\mathfrak{Q}\alpha)(\Psi\alpha \vee \sim\Psi\alpha)$', and no further extension of the rules of inference would be required. Second, we are now in a position to see more clearly what would be signified by the formula '$\Phi\Phi$' if it were admitted. Since the first occurrence of 'Φ' in this formula is one in which it is limited to propositional forms written with a Φ-letter, clearly the second occurrence must be capable of being understood with the same limitation if the formula is to mean anything at all. Hence when we expand the first occurrence by the rule just given, so as to obtain '$(\mathfrak{Q}\alpha)(\Phi\alpha)$', we must still understand the second and remaining occurrence of 'Φ' as limited to propositional forms written with a schematic Φ-letter, so that '$(\mathfrak{Q}\alpha)(\Phi\alpha)$' must here be understood as its special case '$(\mathfrak{Q}\psi)(\Phi\psi)$'. Now in this last formula we may expand the remaining occurrence of 'Φ' in the same way as we expanded the first, and thus we obtain '$(\mathfrak{Q}\psi)(\mathfrak{Q}\alpha)(\psi\alpha)$' as the natural unabbreviated version of '$\Phi\Phi$'. But we have already had to rule that '$(\mathfrak{Q}\psi)(\mathfrak{Q}\alpha)(\psi\alpha)$' is not well-formed, because although there are propositions of this form according to my original explanations, we must not say so if the usual rules of propositional logic are to be retained; all that we are permitted to say is that there are propositions of the form '$(Q\psi)(Q\alpha)(\psi\alpha)$'. One might perhaps say that it is these propositions that are aimed at by the formula '$\Phi\Phi$', but in that case illegitimately, since 'Φ' in this role is introduced as abbreviating '$\mathfrak{Q}$' and not 'Q'.

Now when we consider the introduction of quantifier-symbols in α-position we are met by a complication, though indeed at first glance everything seems very plain sailing. For first, it is perfectly correct to say that any proposition which contains a quantifier somewhere will either be or be built up from a proposition which contains that quantifier initially, and further that a proposition which contains a quantifier initially will always be of the form '$(\mathfrak{Q}\alpha)(\Phi\alpha)$', while a proposition somehow built up from a component of the form '$(\mathfrak{Q}\alpha)(\Phi\alpha)$' must itself be of the related form '$(\mathfrak{Q}_1\phi)(\mathfrak{Q}\alpha)(\phi\alpha)$'. Hence any proposition which contains a quantifier '$\mathfrak{Q}$' must be of the form '$(\mathfrak{Q}_1\phi)(\mathfrak{Q}\alpha)(\phi\alpha)$', and we may perfectly well introduce the formula '$\Phi\mathfrak{Q}$' as a notional variant on this formula. (The new Φ-letter is here being used to abbreviate an original elementary $\mathfrak{Q}\phi$-schema, just as previously it was used to abbreviate an elementary $\mathfrak{Q}\alpha$-schema, and the abbreviating rule required is precisely analogous.) What may possibly seem strange is that if '$\Phi\mathfrak{Q}$' is introduced in this way, then the formulae obtainable from '$\Phi\mathfrak{Q}$' by substituting for 'Φ' can only be those equally obtainable from '$(\mathfrak{Q}_1\phi)(\mathfrak{Q}\alpha)(\phi\alpha)$' by substituting for the elementary $\mathfrak{Q}\phi$-schema '$(\mathfrak{Q}_1\phi)(-\phi-)$', and hence they must be formulae in which '$\mathfrak{Q}$' is used to bind an α-variable. I say this may seem strange, because if '$\Phi\mathfrak{Q}$' is to stand in for any proposition which contains a quantifier '$\mathfrak{Q}$' somewhere, then it must of course stand in for propositions which contain '$\mathfrak{Q}$' used to bind a Φ-variable if '$\mathfrak{Q}$' happens to be a quantifier applicable to Φ-variables. But in fact there is no difficulty here, because of the dummy status of our α-variables: the fact that '$\Phi\mathfrak{Q}$' does *inter alia* represent propositions containing quantifiers binding Φ-variables is catered for by the rule permitting us to go on and substitute Φ-variables for the dummy α-variables bound to '$\mathfrak{Q}$' as the result of our first substitution. Indeed by continued application of the substitution rules stated so far we shall be able to obtain, as substitutions for '$\Phi\mathfrak{Q}$', any of our well-formed formulae containing the letter '$\mathfrak{Q}$' anywhere. And a further point worth remarking is that we shall not now need to give a separate statement of the rule permitting $\mathfrak{Q}$-symbols to be substituted for α-symbols, for it will be found that this rule is a consequence of our extended rule permitting Φ-symbols to be substituted for α-symbols.

A difficulty arises, however, when we consider using *pure* quantifier-symbols in α-position, as in 'ΦQ'. Since 'Q' is in general being treated as a special case of '$\mathfrak{Q}$' (in so far as 'Q' is substitutable for '$\mathfrak{Q}$'), it would seem natural to take 'ϕQ' as a special case of '$\Phi\mathfrak{Q}$', and therefore as an abbreviation for '$(\mathfrak{Q}_1\phi)(Q\alpha)(\phi\alpha)$'. But the result of this would be that some of our well-formed formulae containing 'Q' would *not* be available as substitution-instances of 'ΦQ', which does not seem to be what was intended. As I pointed out, we would in the first instance be able to obtain only those formulae in which 'Q' appeared binding an α-variable, and then with further substitutions some more formulae would become available, but still there would be some that were not forthcoming. (For example, we can obtain '$(Q\phi)(Q\alpha)(\phi\alpha)$' from '$\Phi Q$' by substituting '$Q$' for '$\mathfrak{Q}_1$' in the expansion '$(\mathfrak{Q}_1\phi)(Q\alpha)(\phi\alpha)$', but this method will be available only where 'Q' is schematic; there is no way of obtaining '$(q\phi)(q\alpha)(\phi\alpha)$' from '$\Phi q$' on the present suggestion for interpreting 'Φq'.) Now in fact a great deal can be achieved with this limited understanding of 'ΦQ' as a mere abbreviation—indeed I pointed out on p. 136 that the whole of the present construction of arithmetic could proceed on this basis—but it would be natural to want a more liberal interpretation of the notation, and in that case there seems no alternative to introducing it once more as primitive.

I close this appendix with a brief discussion of how the present logical system fares in the face of the logical paradoxes based on the abstraction principle, for so far as I can see these are all prevented by the restrictions on well-formed formulae that we have already had to introduce for independent reasons. For example the thesis

(i) $(\exists \mathfrak{q}_1)(\forall \mathfrak{q}_2)((\mathfrak{q}_1\phi)(\mathfrak{q}_2\alpha)(\phi\alpha) \equiv \Psi\mathfrak{q}_2)$

is trivially provable from our suggested definition of '$\Psi\mathfrak{q}_2$', and this thesis can be seen as stating that to every predicate of quantifiers Ψ there corresponds some quantifier $\mathfrak{q}_1$, just as the familiar abstraction principle states that to every predicate of objects there corresponds some object. But an attempt to show the inconsistency in our case would have to proceed by substituting for 'Ψ' in (i) to get

(ii) $(\exists \mathfrak{q}_1)(\forall \mathfrak{q}_2)((\mathfrak{q}_1\phi)(\mathfrak{q}_2\alpha)(\phi\alpha) \equiv {\sim}(\mathfrak{q}_2\phi)(\mathfrak{q}_2\alpha)(\phi\alpha))$,

and this step is illegitimate. The clause '${\sim}(\mathfrak{q}_2\phi)(\mathfrak{q}_2\alpha)(\phi\alpha)$' which is introduced in place of 'Ψq_2' is simply ill-formed, and the substitution is not licensed by any of the rules I have given. A precisely similar fate awaits the similar principle

(iii) $(\exists \mathbf{q}_1)(\forall \mathbf{q}_2)((\mathbf{q}_1 \phi)(\phi \mathbf{q}_2) \equiv \Psi \mathbf{q}_2)$,

which—as we shall see in a moment—is easily proved. For again to obtain a contradiction from this we must substitute for 'Ψ' to get

(iv) $(\exists \mathbf{q}_1)(\forall \mathbf{q}_2)((\mathbf{q}_1 \phi)(\phi \mathbf{q}_2) \equiv \sim(\mathbf{q}_2 \phi)(\phi \mathbf{q}_2))$,

and again the substitution is illegitimate. The clause '$(\mathbf{q}_2 \phi)(\phi \mathbf{q}_2)$' expands according to the definition given to '$(\mathbf{q}_2\, \mathbf{q}_3)(\mathbf{q}_3\, \phi)(\mathbf{q}_2\, \alpha)(\phi\alpha)$', and this is just as ill-formed as the last.

Of course the crucial but ill-formed clauses in these deductions would have been well-formed if in place of the general quantifier-variable '$\mathbf{q}_2$' we had written the pure quantifier-variable 'q_2', and it is worth seeing what happens to the arguments if we rephrase them in this way. Assuming, then, that the notation with pure quantifier-symbols in α-position is introduced as primitive, there will be no way of establishing the leading principle analogous to (i), namely

$$(\exists \mathbf{q}_1)(\forall q_2)((\mathbf{q}_1 \phi)(q_2\, \alpha)(\phi\alpha) \equiv \psi q_2),$$

because (i) rested on the *definition* of '$\Psi \mathbf{q}_2$'. But we can still establish the principle analogous to (iii), by starting with the trivial theorem

(v) $(\exists \mathbf{q}_1)(\forall q_2)((\mathbf{q}_1 \phi)(\phi q_2) \equiv (\mathfrak{Q}\phi)(\phi q_2))$,

and then substituting for the elementary $\mathfrak{Q}\phi$-schema '$(\mathfrak{Q}\phi)(-\phi-)$' the further $\mathfrak{Q}\phi$-schema '$(-\Psi-)$' to obtain

(vi) $(\exists \mathbf{q}_1)(\forall q_2)((\mathbf{q}_1 \phi)(\phi q_2) \equiv \Psi q_2)$.

Then if 'Ψq_2' does indeed range over all contexts for 'q_2' we may quite properly infer

(vii) $(\exists \mathbf{q}_1)(\forall q_2)((\mathbf{q}_1 \phi)(\phi q_2) \equiv \sim(q_2\, \phi)(\phi q_2))$.

But no contradiction results from this, since the two quantifier-variables involved are now of different types. What *would* yield a contradiction would be

(vii)′ $(\exists q_1)(\forall q_2)((q_1\, \phi)(\phi q_2) \equiv \sim(q_2\, \phi)(\phi q_2))$

and this could be validly obtained from

(vi)′ $(\exists q_1)(\forall q_2)((q_1\, \phi)(\phi q_2) \equiv \Psi q_2)$.

But there is no way of deducing this last. In particular it does *not* follow from the simple theorem

(v)′ $(\exists q_1)(\forall q_2)((q_1\, \phi)(\phi q_2) \equiv (Q\phi)(\phi q_2))$,

since the schema '$(-\Psi-)$', though a perfectly good $\mathfrak{Q}\phi$-schema, is *not* a pure Q-schema.

5. Numbers

1. Numerical Quantifiers and Identity

I have said that numerical quantifiers can be applied in all types, and this would seem to imply that numerical quantifiers must be construed as entirely pure and unrestricted quantifiers which are applicable to any monadic propositional form without further ado. Now the numerical quantifiers we started off by considering were defined in terms of the universal quantifier and *identity*, and we have already seen that the universal quantifier is totally unrestricted; but can the same be maintained for the notion of identity? Does identity apply unchanged in any type whatever?

As I have remarked, Frege's view seems to have been that identity could hold only between objects, or in other words that the identity sign was capable of occurring in singular statements only between expressions whose grammatical role is that they are used to mention or refer, and in other statements only if they are got from these statements by generalization. This view would certainly seem to be borne out by ordinary usage. For instance if one wanted to claim *identity* between the predicates expressed by '. . . is a man' and '. . . is human' one might perhaps offer the version

> To say that anything is a man is the same as to say that it is human

which is plainly a generalization over such propositions as

> To say that Socrates is a man is the same as to say that Socrates is human

and it seems clear that the expressions here flanking '. . . is the same as . . .' are *nominalizations* of the propositional expressions 'Socrates is a man' and 'Socrates is human', and therefore (grammatical) *subject*-expressions. And the same is of course true of the expressions I employed initially, namely 'the predicate expressed by ". . . is a man" ', and 'the predicate expressed by ". . . is human" '. So far as our ordinary usage goes, then, Frege's view would appear to be sound, and rather than combat it directly I propose to start by arguing that it is really irrelevant to our purpose, for the connection between numerical quantifiers and identity is not at all what Frege supposed.

In ordinary discourse the quantifier expressed by 'there are three . . .' is generally expressed equally well by the apparently pleonastic locution 'there are three *different* . . .'. Thus, 'there are three men in the room' is clearly interchangeable with 'there are three different men in the room', where this last may be interpreted as 'there are three men in the room such that each is a different man from the other two'. The locution is pleonastic because of course it must be true that each man is a different man from all *other* men, i.e. from all men that are different men from him, but I think it is interesting to notice that we do at least have this locution, for it serves to emphasize that what one might call the 'counting relation' here *is* the relation of *difference* or non-identity. And the significance of this is precisely that it is not always the case that the counting relation is non-identity—indeed this is not always the case even when what are being counted are such ordinary objects as men.

For consider the proposition

> There are at least three men of different height in the room.

If we try to analyse this in terms of the familiar numerical quantifiers employed hitherto, we might start by putting it in the form

> There are at least three men in the room such that each of them is of a different height from the other two

but to complete our analysis we now have to consider the phrase 'the other two', which is plainly to be understood in terms of the three we began with. For it will not do to drop the word 'two' so as to obtain '. . . each is of a different height from *all* the other (men in the room)' or '. . . each is of a different height from *some* other (men in the room)', since in the first case our proposition would be false if there were four men in the room and two of them were of the same height, while in the second case it would be true if there were three men in the room and two of them were of the same height, and neither of these is what we intended. Nor will it do to retain the word 'two' but to drop the word 'the' so as to obtain '. . . each is of a different height from *every* other two (men in the room)' or '. . . each is of a different height from *some* other two (men in the room)', for in the first case our proposition would be false if there were five men in the room and three of them were of the same height, while in the second case our proposition would be true if there were six men in the room but three of them were of one height and three of another, and again neither of these is what we intended. So it would seem that we cannot achieve what we want by taking the phrase 'the other two' as concerning merely the other *men in the room*; it must be taken more narrowly as concerning the others *of the three we started with*, and this introduces some complication.

It may seem that a solution to the difficulty could be obtained by a slight modification of the last suggestion, in which 'the other two' was replaced by 'some other two'. For this suggestion broke down only because the other two in question might be of the same height as each other, and so we could put matters right by adding to our specification that each is to be 'of a different height from some (other) two men in the room *of different height*'. This would then yield the analysis

> There are at least three men in the room such that for each of them *there are at least two men of different height in the room* such that each of them (the latter) is of a different height from him (the former).

But the italicized portion of this analysis contains the very notion we were trying to analyse all over again, though this time with 'two' in place of 'three'. Repeating the analysis (somewhat pedantically) for the italicized portion, and introducing the abbreviations 'Fx' for 'x is a man in the room' and 'x≈y' for 'x is the same height as y', we find in the end that our original proposition may be written

$$(\exists 3x)(Fx \;\&\; (\exists 2y)(Fy \;\&\; y \not\approx x \;\&\; (\exists 1z)(Fz \;\&\; z \not\approx x \;\&\; z \not\approx y))).$$

This pattern of analysis is applicable to any proposition of the form 'there are n so-and-so's of different height', but I think it should be clear that it is a cheat. Although the analysis will begin with the familiar numerical quantifier '(∃nx)(—x—)', we cannot regard this as showing that our original proposition splits into the parts 'there are n . . .' and '. . . so-and-so's of different height', for *what* we are to take the quantifier 'there are n. . .' as being applied to will partly depend upon *which* numerical quantifier this is, and it certainly cannot be determined by reflecting on the alleged part '. . . so-and-so's of different height'. And a very strong indication that the whole analysis is a fabrication is that the initial numerical quantifier of the analysis, and indeed every other numerical quantifier in it as well, is *entirely superfluous,* and may be replaced without loss by a simple existential quantifier. Given only the information that '... ≈ ...' is a reflexive relation, the formula written above can very easily be shown to be equivalent to

$$(\exists x)(Fx \; \& \; (\exists y)(Fy \; \& \; y \not\approx x \; \& \; (\exists z)(Fz \; \& \; z \not\approx y \; \& \; z \not\approx x))).$$

The moral that I want to draw from this is that a proposition such as 'there are at least three men of different height in the room' should be regarded rather as containing a part expressed by '. . . men . . . in the room' to which there is applied a quantifier, here expressed by 'there are at least three . . . of different height', but perhaps more perspicuously written as 'there are at least three *differently-heighted* . . .' so as to provide the right grammatical contrast with the familiar quantifier 'there are at least three *different* . . .'. Whereas successive instances of the familiar numerical quantifiers are defined by the rule

$$(\exists n' \text{ different } x)(Fx) \equiv (\exists x)(Fx \; \& \; (\exists n \text{ different } y)(Fy \; \& \; y \text{ differs from } x))$$

successive instances of the new quantifiers are equally given by the rule

$$(\exists n' \text{ differently-heighted } x)(Fx) \equiv (\exists x)(Fx \; \& \; (\exists n \text{ differently-heighted } y)\,(Fy \; \& \; y \text{ differs in height from } x)).$$

And I shall say that in the familiar case the 'counting relation' is the relation '. . . differs from . . .', while in the second case the 'counting relation' is rather the relation '. . . differs in height from . . .'.

Now the relation '. . . differs in height from . . .' may seem to be acceptable as a counting relation only because it shares with the paradigm counting relation '... differs from ...' the property of being the negation of an equivalence relation.[1] However, it seems that in fact we do not restrict ourselves to using only relations with this property as counting relations. For instance the statement

> There are at least three mutually exclusive classes which have four members apiece chosen from among the men in this room

(which is a longwinded way of saying that there are at least 3×4 men in this room) is very naturally regarded as opening with the numerical quantifier 'there are at least three *mutually exclusive* . . .'. Using 'α' and 'β' temporarily as class-variables, and defining

$$`\alpha \mid \beta' \quad \text{for} \quad `{\sim}(\exists x)(x \in \alpha \mathbin{\&} x \in \beta)'$$

we may say that these quantifiers are given by the rule

$$(\exists n' \text{ exclusive } \alpha)(\Phi\alpha) \equiv (\exists \alpha)(\Phi\alpha \mathbin{\&} (\exists n \text{ exclusive } \beta)(\Phi\beta \mathbin{\&} \beta \mid \alpha))$$

and it is easy to see that this would give the right results here. But the counting relation '... | ...' (so defined) is not by any means the negation of an equivalence relation.

However, I think it would be rash to generalize from these two examples to the very sweeping conclusion that any relation whatever may plausibly be taken as a counting relation. First, there seems an evident absurdity in taking *identity* itself as a counting relation (so that one could, for instance, raise the question of how many *identical* men there were in the room), for —to bring just one argument—it is undoubtedly part of our understanding of numerical quantifiers that 'there are two . . .' and 'there are three . . .' have different truth-conditions, and yet it would have to be held that the quantifiers 'there are n *identical* . . .' all have exactly the same truth-conditions, no matter what numeral is put in place of 'n'. A moment's reflection shows that this same argument holds for any other reflexive relation that we might consider in place of identity, and further that a similarly unacceptable result is liable to arise for any relation which is not actually irreflexive. For if there is even one object in a collection which bears our relation to itself, then it follows that if we are counting the collection with that relation as our

[1] A relation is an equivalence relation if and only if it is both transitive and symmetrical (and therefore, within its field, reflexive).

counting relation *all* the quantifiers 'there are at least n...' will be applicable with equal correctness, which again seems quite contrary to our normal understanding (at least when the collection is finite). So the first restriction to impose seems to be that a relation is intuitively acceptable as a counting relation only if it is irreflexive.

It is tempting to put this requirement by saying that any acceptable counting relation must at least be contained in non-identity, thus preserving a central place for non-identity as the paradigm counting relation, but at this point we should introduce a pertinent distinction. Certainly it is easily shown that a relation '... R ...' between objects is irreflexive if and only if it satisfies the condition

$$(\forall x)(\forall y)(xRy \supset \sim(\forall \mathfrak{f})(\mathfrak{f}x \equiv \mathfrak{f}y))$$

but, even when we are confining our attention to relations between objects, it may be disputed whether the Leibnizian condition for non-identity which figures as consequent to this implication is after all acceptable as a condition for non-identity in the ordinary sense. One ground for doubt would be that the quantifier '$(\forall \mathfrak{f})(— \mathfrak{f} —)$' presumably includes intensional properties within its range—at least, we have not stipulated that it shall not—and there is no reason to suppose that identical objects must share all their *intensional* properties.[2] Another (but more controversial) ground would be that in our ordinary everyday way of thinking there is really no such thing as *the* relation of identity; rather, there are various different identity-relations exemplified, for instance, by the relations of being *the same man*, of being *the same river*, or of being *the same sentence*, and none of these can be guaranteed to carry with it even the sharing of all extensional properties.[3] Now we do not need to investigate these doubts in any detail, but they do show that it would be unwise to assume without more ado that the ordinary notion of identity coincides

[2] Whether the quantifier '$(\forall \mathfrak{f})(— \mathfrak{f} —)$' ranges over intensional properties depends upon whether there are propositions of the form 'Fa' in which 'F...' is an intensional predicate, and therefore it depends upon whether we must recognize propositions with apparently intensional predicates as genuinely *containing a subject*, or can instead offer to 'eliminate' their apparent subject-expressions. Compare chapter 2, section 3.

[3] For instance it will be argued that it may be true that x is the same river as y, false that x consists of the same water as y, but true that x consists of the same water as x. See P. T. Geach, *Reference and Generality*, §§ 91–2.

with that defined by Leibniz's condition. And my reason for introducing this point here is of course that it is only Leibnizian identity that has a claim to be recognized as fundamental for the notion of a numerical quantifier. We could perhaps put this claim by saying that one should never count things as *two* unless there is *some* difference between them, but in that case we should certainly leave it as a possibility that even identical objects might turn out to exhibit *some* difference, at least so far as our ordinary notion of identity is concerned.

When, therefore, our concern is with the numerical quantifiers we may reasonably ignore any arguments based on the ordinary notion of identity, and confine our attention to the identity given by Leibniz's condition. And so, to revert for a moment to the starting point of this discussion, we are now in a position to reject the argument I put forward initially on Frege's behalf. In the first place it is not true that all numerical quantifiers must be defined by reference to identity, and secondly in so far as the notion of identity does have a fundamental role to play in the definition of numerical quantifiers it is Leibnizian identity that we require, and Leibnizian identity does apply unchanged *in any type whatever*. Leibnizian identity (which is what I shall henceforth mean by identity) can be evidently be defined by the type-neutral definition

$$\text{`}A=B\text{'} \quad \text{for} \quad \text{`}(\forall\phi)(\phi A \equiv \phi B)\text{'}.$$

When A and B are taken as subjects or as predicates, then this definition will lead us to the equivalences[4]

$$a=b \equiv (\forall\mathfrak{f})(\mathfrak{f}a \equiv \mathfrak{f}b)$$

$$F=G \equiv (\forall\mathfrak{q})((\mathfrak{q}x)(Fx) \equiv (\mathfrak{q}x)(Gx))$$

while if A and B are taken as pure quantifiers we have simply

$$Q_1=Q_2 \equiv (\forall\phi)(\phi Q_1 \equiv \phi Q_2)$$

but all cases can perfectly well be viewed as consequences of the same type-neutral definition, so there seems no reason to deny that it is the same notion of (Leibnizian) identity that occurs in all of them.

To return to my theme, I have suggested so far that a relation will be intuitively acceptable as a counting relation only if it is

[4] This is an application of my remarks on pp. 125–6 concerning the interpretation of 'Φa' and 'ΦF'.

irreflexive, and hence contained in (Leibnizian) non-identity. This does appear to be the most important requirement on counting relations, but it also seems reasonable to impose a second requirement, namely that the relation be symmetrical. The reason for this is that if we attempt to count a set of things under a counting relation that is not symmetrical then the result that we reach will generally depend upon the *order* in which we choose to count the set, and this seems quite inappropriate. It is true that the present custom is to speak of 'ordinal' numbers as well as 'cardinal' numbers, and ordinal numbers are numbers whose application depends upon the order in which things are considered, but I think one may well feel hesitant about allowing these ordinal numbers to be *numbers* in any but a very recondite sense, and besides the distinction is generally employed only where it is infinite sets that are in question, and therefore where the simple numerical quantifiers that we are considering are ineffective anyway. So, as I say, it seems to me that we may very reasonably require an acceptable counting relation to be symmetrical as well as irreflexive, but I cannot see any other requirement that it would be proper to impose on all cases. To sum up these requirements, then, let us set down the definition[5]

'Count ($\mathfrak{R}$)' for '$(\forall\alpha)(\sim\alpha\mathfrak{R}\alpha)\ \&\ (\forall\alpha)(\forall\beta)(\alpha\mathfrak{R}\beta \equiv \beta\mathfrak{R}\alpha)$'.

Now for the general form of a weak numerical quantifier I propose to adopt the schema '$(\exists n_{\mathfrak{R}}\alpha)(—\alpha—)$', where '$\mathfrak{R}$' represents the counting relation involved. (The whole schema may be read 'There are at least n α's, counted under $\mathfrak{R}$, such that —α—'.) My argument so far has been that a proposition does not count as a satisfactory proposition of the form '$(\exists n_{\mathfrak{R}}\alpha)(\Phi\alpha)$' unless it is also true that Count ($\mathfrak{R}$), but for formal purposes we have to make up our minds what to say about a proposition '$(\exists n_{\mathfrak{R}}\alpha)(\Phi\alpha)$' where this condition is not satisfied, for we can hardly be prevented from framing such propositions. One possibility would be to ensure that they were all *false* by adding the clause 'Count ($\mathfrak{R}$)' into our definition of the formula '$(\exists n_{\mathfrak{R}}\alpha)(\Phi\alpha)$', but it is slightly more convenient from a technical point of view to give a definition of this formula which allows

[5] I use the new letter '$\mathfrak{R}$' in place of 'R' for the same reason that I use 'Φ' in place of 'F', namely as a schematic letter for relations of *any* type, and not only for relations between objects. But '$\mathfrak{R}$' will always be used for a *homogeneous* relation, i.e. a relation whose terms are both of the same type.

such propositions to be on occasion true, though quite uninteresting. That is, we give a general definition of the formula '$(\exists n_{\mathfrak{R}}\alpha)(\Phi\alpha)$' which does not itself limit '$\mathfrak{R}$' to relations such that Count ($\mathfrak{R}$), but we shall only be concerned to ensure the intuitive adequacy of this definition in cases where the condition is satisfied.

To help us in framing this definition, let us consider once more the proposition that there are at least three men of different height in the room. We have said, in effect, that this can be equivalently put in the form

> There are at least three men in the room such that each of those three is of a different height from the others of those three

and the problem is that we have to single out from all the men in the room some three (non-identical) men who are each of a different height from one another. Using the same abbreviations as previously we therefore require that there should be objects a, b, c such that

(i) $a \neq b \ \& \ a \neq c \ \& \ b \neq c$

(ii) $a \not\approx b \ \& \ a \not\approx c \ \& \ b \not\approx c$

(iii) $Fa \ \& \ Fb \ \& \ Fc$.

Now we can represent all these three conditions by making use of the same one-place predicate '$\ldots = a \vee \ldots = b \vee \ldots = c$'. For the first condition is easily shown to be equivalent to

$$(\exists 3x)(x=a \vee x=b \vee x=c)$$

(where the numerical quantifier is one of our initial numerical quantifiers based on non-identity); similarly the second condition can equally well be put as

$$(\forall x)(\forall y)((x=a \vee x=b \vee x=c) \ \& \ (y=a \vee y=b \vee y=c) \ \& \ x \neq y \supset x \not\approx y)$$

(for the relation '$\ldots \not\approx \ldots$' is of course symmetrical); while finally the third condition is evidently the same as

$$(\forall x)((x=a \vee x=b \vee x=c) \supset Fx).$$

We see, then, that if there are three men in the room of different height then there is some true proposition of the form (with respect to 'G')

$$(\exists 3x)(Gx) \ \& \ (\forall x)(\forall y)(Gx \ \& \ Gy \ \& \ x \neq y \supset x \not\approx y) \ \& \ (\forall x)(Gx \supset Fx)$$

or in other words we see that

$$(\exists \mathfrak{g})((\exists 3x)(\mathfrak{g}x) \;\&\; (\forall x)(\forall y)(\mathfrak{g}x \;\&\; \mathfrak{g}y \;\&\; x \neq y \supset x \not\approx y) \;\&\; (\forall x)(\mathfrak{g}x \supset Fx)).$$

Further, a moment's reflection suffices to show that the converse is also true, and that the same pattern of reasoning will be applicable to any proposition which contains a similar numerical quantifier. So we are led to the definition

$$\text{`}(\exists n_{\mathfrak{R}}\,\alpha)(-\alpha-)\text{' for `}(\exists \psi)((\exists n\alpha)(\psi\alpha) \;\&\; (\forall\alpha)(\forall\beta)(\psi\alpha \;\&\; \psi\beta \;\&\; \alpha\neq\beta \supset \alpha\mathfrak{R}\beta) = (\forall\alpha)(\psi\alpha \supset -\alpha-))\text{'}$$

which I forthwith adopt.[6]

The quantifier '$(\exists n\alpha)(-\alpha-)$' which appears in this definiens is, as I say, intended as a quantifier based on non-identity as its counting relation, but I have not written it in the form '$(\exists n_{\neq}\,\alpha)(-\alpha-)$' because of course *that* form is being introduced by this definition itself, as short for

$$(\exists \psi)((\exists n\alpha)(\psi\alpha) \;\&\; (\forall\alpha)(\forall\beta)(\psi\alpha \;\&\; \psi\beta \;\&\; \alpha\neq\beta \supset \alpha\neq\beta) \;\&\; (\forall\alpha)(\psi\alpha \supset -\alpha-)).$$

It is evident that the two quantifiers '$(\exists n\alpha)(-\alpha-)$' and '$(\exists n_{\neq}\,\alpha)(-\alpha-)$' will coincide,[7] and I shall in fact employ the two notations interchangeably in what follows, but for formal purposes we should retain them both if we are not to have the same symbol furnished with two distinct definitions. For of course although I have just defined the general form of a numerical quantifier '$(\exists n_{\mathfrak{R}}\,\alpha)(-\alpha-)$' in terms of what is really its special case '$(\exists n\alpha)(-\alpha-)$', the special case itself still awaits definition. But before I proceed to this I should pause to give some elaboration of the definition just adopted.

The quantifier-expression '$\exists n_{\mathfrak{R}}$' is defined to mean, roughly, 'There are, when counted under $\mathfrak{R}$, at least n . . .', and it is perhaps not clear whether this definition carries with it analogous

[6] We could generalize this definition to deal with cases where '$\mathfrak{R}$' is more than dyadic, as in the example 'there are at least n points, no three of which are colinear, such that . . .'. But I here confine myself to dyadic relations, partly because it is only dyadic relations that will be used hereafter, and partly because I suspect (as I hinted on pp. 147–8) that our *usual* counting relations are dyadic relations which we express with the words 'not the same' but which are not to be identified with Leibnizian non-identity.

[7] A formal proof is provided in Appendix 2, Theorem 0.1.

definitions of, e.g. 'there are, when counted under $\mathfrak{R}$, at least 3...' or 'there are, when counted under $\mathfrak{R}$, at least n+1 ...'. To avoid any doubts, then, I now lay down the more general definition

$$\text{`}(Q_{\mathfrak{R}}\alpha)(\text{—}\alpha\text{—})\text{'} \quad \text{for} \quad \text{`}(\exists\psi)((Q\alpha)(\psi\alpha)\ \&\ (\forall\alpha)(\forall\beta)(\psi\alpha\ \&\ \psi\beta\ \&\ \alpha\neq\beta \supset \alpha\mathfrak{R}\beta)\ \&\ (\forall\alpha)(\psi\alpha \supset \text{—}\alpha\text{–})),$$

which is to apply to any pure quantifier-expression in place of 'Q'—though of course it will be interesting only when in place of 'Q' we do have a constant or schematic expression for a (weak) numerical quantifier. (The whole expression '$Q_{\mathfrak{R}}$' will also stand for a quantifier, but not necessarily for a *pure* quantifier, since '$\mathfrak{R}$' may be so chosen as to be restricted to this or that type.) By this means we do obtain definitions of '$(\exists 3_{\mathfrak{R}}\alpha)(\text{—}\alpha\text{—})$' and of '$((\exists n')_{\mathfrak{R}}\alpha)\ (\text{—}\alpha\text{—})$' as desired—for the expression '$\exists n'$' will shortly be defined as a pure quantifier expression. In this last case I shall generally omit the cumbersome internal brackets, so as to write simply '$(\exists n'_{\mathfrak{R}}\alpha)(\text{—}\alpha\text{—})$',[8] and this provokes a further reflection. For one might *also* expect to find '$(\exists n'_{\mathfrak{R}}\alpha)(\text{—}\alpha\text{—})$' defined directly in terms of '$(\exists n_{\mathfrak{R}}\alpha)(\text{—}\alpha\text{—})$', so that we could set down as a matter of definition the equivalence

$$\vdash (\exists n'_{\mathfrak{R}}\alpha)(\Phi\alpha) \equiv (\exists\alpha)(\Phi\alpha\ \&\ (\exists n_{\mathfrak{R}}\beta)(\Phi\beta\ \&\ \beta\mathfrak{R}\alpha)).$$

But of course it does not matter if this thesis is not forthcoming as a trivial matter of definition, so long as it is in fact forthcoming; and indeed it *is* forthcoming, *but* only on the condition that Count ($\mathfrak{R}$). Upon this condition it is possible to prove from our definitions both of the recursive equivalences[9]

$$\vdash (\exists 1_{\mathfrak{R}}\alpha)(\Phi\alpha) \equiv (\exists\alpha)(\Phi\alpha)$$

$$\vdash (\exists n'_{\mathfrak{R}}\alpha)(\Phi\alpha) \equiv (\exists\alpha)(\Phi\alpha\ \&\ (\exists n_{\mathfrak{R}}\beta)(\Phi\beta\ \&\ \beta\mathfrak{R}\alpha))$$

and this is what I had in mind when I said earlier that we should only expect our definition to be intuitively adequate when this condition is satisfied. For I think it very reasonable to take the availability of these two recursive equivalences as showing that the definition is indeed adequate to our intuitive understanding of the notion of a numerical quantifier.

There will in fact be few theorems of any interest which hold

[8] Note that '$(\exists n'_{\mathfrak{R}}\alpha)(\Phi\alpha)$' abbreviates '$((\exists n')_{\mathfrak{R}}\alpha)(\Phi\alpha)$, and not—what would be quite different—'$((\exists n_{\mathfrak{R}})'\alpha)(\Phi\alpha)$'.

[9] See Appendix 2, Theorems 0.3–0.5.

for *all* numerical quantifiers but do not require the condition 'Count ($\mathfrak{R}$)', and incidentally there will be several important theorems which require even stronger conditions. For instance the theorem which I demonstrated as my second lemma in the appendix to chapter 1 was there written simply as

$$(\exists n'x)(Fx) \supset (\exists nx)(Fx \ \& \ x \neq a)$$

but if we now consider the corresponding but more general thesis

$$(\exists n'_{\mathfrak{R}}\alpha)(\Phi\alpha) \supset (\exists n_{\mathfrak{R}}\alpha)(\Phi\alpha \ \& \ \alpha\mathfrak{R}\beta)$$

we find that the more general thesis will hold whenever '... $\mathfrak{R}$ ...' is taken as the negation of an equivalence relation, as a glance at the proof will show, but it will fail even for the relation of being mutually exclusive, which we have already seen to be a perfectly acceptable counting relation.[10]

In conclusion, then, I have adopted the schema '$(\exists n_{\mathfrak{R}}\alpha)$ $(—\alpha—)$' to represent the general form of a weak numerical quantifier, and I permit any relation whatever to take the place of '$\mathfrak{R}$' in this schema. But I have also adopted a definition of this general form which in effect defines it in terms of its special case '$(\exists n_{\neq}\alpha)(—\alpha—)$', and so makes the quantifier uninteresting unless '... $\mathfrak{R}$ ...' is taken as a symmetrical and irreflexive relation. The adoption of this definition would seem to conform to our natural assumption that non-identity is the central case of a counting relation, and it ensures that we can determine the properties of all numerical quantifiers—and hence establish the foundations of arithmetic—so long as we can determine the properties of these central numerical quantifiers. The task before us is, then, just to introduce the numerical quantifiers based on non-identity, which is indeed precisely the task that we had envisaged in the first place. But the present discussion has shown that any objection to the programme which argues from the nature of the identity relation must miss the mark.

[10] It may be noted that the thesis will hold with '...|...' in place of '...$\mathfrak{R}$...' if we add the condition 'ΦB' and choose 'Φ...' appropriately, in particular if we choose 'Φ...' so that it includes a conjunct of the form '... has just n members' as it did in my original example. The effect of this choice of 'Φ...' is to limit our attention to classes with the same number of members, and I think it is true that the quantifiers 'there are n mutually exclusive classes which ...' are most naturally used only with this limitation. But even with this limitation the counting relation '...|...' is still not the negation of an equivalence relation.

2. The Introduction of Numerical Quantifiers

So, when exactly is a proposition of the form '$(\exists n\alpha)(\Phi\alpha)$'? Well, it is clear that every proposition of the form '$(\exists\alpha)(\Phi\alpha)$' is also of the form '$(\exists n\alpha)(\Phi\alpha)$', and that wherever a proposition is of the form '$(\exists n\alpha)(\Phi\alpha)$' then the proposition of the related form '$(\exists\alpha)(\Phi\alpha \;\&\; (\exists n\beta)(\Phi\beta \;\&\; \beta\neq\alpha))$' is also of the form '$(\exists n\alpha)(\Phi\alpha)$', but I think that these two rules may not be enough to give us a complete characterization of the form '$(\exists n\alpha)(\Phi\alpha)$'. To enable us to speak more concisely here, let us suppose that we already have available an infinite string of explicit definitions:

'$(\exists 1\alpha)(\Phi\alpha)$' for '$(\exists\alpha)(\Phi\alpha)$'

'$(\exists 2\alpha)(\Phi\alpha)$' for '$(\exists\alpha)(\Phi\alpha \;\&\; (\exists 1\beta)(\Phi\beta \;\&\; \beta\neq\alpha))$'

'$(\exists 3\alpha)(\Phi\alpha)$' for '$(\exists\alpha)(\Phi\alpha \;\&\; (\exists 2\beta)(\Phi\beta \;\&\; \beta\neq\alpha))$'

etc.

Then the two rules we have stated enable us to say that if a proposition is of one of these explicitly defined forms then it will be of the form '$(\exists n\alpha)(\Phi\alpha)$', but we ought to consider the objection that there may be propositions which can quite properly be said to be of the form '$(\exists n\alpha)(\Phi\alpha)$' even though they are not to be analysed as of one of these explicitly defined forms (or, as I shall also say, even though they do not contain one of these explicitly defined quantifiers initially). Roughly speaking, the objection would be that these propositions could not be analysed as of some form '$(\exists 1''\cdots'\alpha)(\Phi\alpha)$' either because they ought to be analysed differently or because they ought to be admitted to be unanalysable in the relevant respect.

As an example of the first sort we might consider a proposition expressed by means of the English expression 'There are three hundred . . .', and as an example of the second we might recur to the supposedly unanalysable quantifier 'there are V . . .'. In the first case the objection would be that the *meaning* of the English expression 'three hundred' is clearly to be explained in terms of the meanings of 'three' and 'hundred', and not in terms of the meaning of 'one more than two hundred and ninety-nine', and therefore the quantifier 'there are three hundred . . .' does not have the same meaning as any of our explicitly defined quantifiers. And in the second case the objection is simply that the relevant explicitly defined quantifier '$(\exists 5\alpha)(-\alpha-)$'—or. as the case may be, '$(5\alpha)(—\alpha—)$'— is complex in meaning and defined by its position in a series, while 'there are V . . .' is by hypothesis entirely simple in meaning and quite independent of any series.

Now it does not matter very much whether we insist that these examples *can* after all be analysed as one of our explicitly defined quantifiers, or whether we prefer to admit that although they are indeed numerical quantifiers they are not any of the explicitly defined quantifiers. In the first case we would, technically speaking, be regarding our schematic letters for numerals as standing only in the place of the explicitly defined quantifiers, but it would then be necessary for us to elucidate further why it is that our awkward quantifiers may be 'analysed as' the relevant explicitly defined quantifiers. In the second case we would be allowing that the schematic letters can take the place of other quantifiers than those explicitly defined, and it would be necessary to specify which these other quantifiers are. But in each case the problem to be faced is the same, namely: what is the relation between the explicitly defined quantifiers and such quantifiers as 'there are three hundred . . .' and 'there are V . . .' which entitles us to say that these latter are indeed numerical quantifiers of the relevant sort? Whether we call this desired relation the relation of being 'analysable as' would seem to be largely a verbal point of no importance.

Now it seems to me that this problem is adequately resolved if we specify that a proposition is of the form '$(\exists n\alpha)(\Phi\alpha)$' if and only if it contains initially a *pure quantifier* which is *equivalent to* one of the explicitly defined quantifiers, or in other words that a proposition is of the form '$(\exists n\alpha)(\Phi\alpha)$' if and only if it is of the form '$(Q\alpha)(\Phi\alpha)$' where there is a true proposition of one of the related forms

$$(\forall\phi)((Q\alpha)(\phi\alpha) \equiv (\exists 1\alpha)(\phi\alpha))$$

or
$$(\forall\phi)((Q\alpha)(\phi\alpha) \equiv (\exists 2\alpha)(\phi\alpha))$$

or
$$(\forall\phi)((Q\alpha)(\phi\alpha) \equiv (\exists 3\alpha)(\phi\alpha))$$

and so on. One might perhaps wonder whether a mere material equivalence is a strong enough relation for this purpose, or whether perhaps a necessary equivalence (or even an identity) would be needed instead. But I think that this doubt vanishes on closer reflection.

It does very often happen that a mere material equivalence is not a strong enough relation to use for purposes of analysis, but this is mainly because a material equivalence may hold as it were *accidentally*, simply on the ground that both equivalents are true of nothing; however this mishap can hardly happen in our case. For the explicitly defined quantifiers are, as we have insisted, *pure* quantifiers applicable unchanged to every monadic form without exception, and from this it is easy to see that each one of these quantifiers must have some true applications (and some false ones). One relevant consideration[11] is that there is an unending supply of distinct monadic forms taken from the orthodox hierarchy of types, but we do not need to step outside the present logical system to establish the point, for it is easily seen from Frege's theorem (that there are just n numbers less than or equal to n) that the numerical quantifiers themselves can provide all the monadic forms required. If, then, each explicitly defined numerical quantifier must have some true applications, then any *pure* quantifier that is equivalent to an explicitly defined numerical quantifier must also have those same true applications, and the criterion I offer cannot be trivialized in this way.

It is important to observe, though, that the success of this criterion depends upon the fact that we have already restricted our attention to *pure* quantifiers. For suppose that we had merely said that for a proposition to be of the form '$(\exists n\alpha)(\Phi\alpha)$' all that was required was that it contain initially some quantifier or other that is equivalent (in the sense displayed) to one of the explicitly defined numerical quantifiers. It would then follow that some type-restricted quantifiers would satisfy the criterion, for indeed a subject-restricted numerical quantifier such as '$(\exists 3x)(—x—)$' must be equivalent in the relevant sense to the pure quantifier '$(\exists 3\alpha)(—\alpha—)$', and similarly with any other numeral in place of '3'. This is because to say that these two quantifiers are equivalent in the sense in question is just to say that all propositions of the form

$$(\exists 3x)(\Phi x) \equiv (\exists 3\alpha)(\Phi\alpha)$$

are true, which indeed they must be. For in order to obtain a proposition of this form we must write the same monadic

[11] Which, incidentally, is the basis of the remark in note 15 to chapter 4.

propositional form in place of both occurrences of the letter 'Φ', and in order that our left hand equivalent be a proposition we must in fact choose a monadic form which is written with a subject-letter as its schematic letter; consequently the only propositions of this form will be propositions that are more particularly of the form

$$(\exists 3x)(\Phi x) \equiv (\exists 3x)(\Phi x)$$

—or, as it is more usually written,

$$(\exists 3x)(Fx) \equiv (\exists 3x)(Fx).$$

Hence, the two quantifiers '$(\exists 3x)(—x—)$' and '$(\exists 3\alpha)(—\alpha—)$' are indeed equivalent in the relevant sense.

But now, equivalence with an explicitly defined but type-restricted numerical quantifier would *not* appear to be a strong enough criterion here, because it would seem possible that this criterion *could* be satisfied accidentally. Since we are not construing numbers themselves as objects, and since we have been adopting a sceptical attitude to those purported objects that are described by nominalizations of predicate-expressions and the like, we should certainly admit the possibility of there being only a finite number of objects. But if in fact there were, say only 99 objects, then the quantifier '$(\exists 100x)(—x—)$' would have no true applications, and so would be equivalent to every other subject-quantifier that had no true applications, for instance to the quantifier '$(\forall x)(—x— \,\&\, \sim —x—)$'. Hence a proposition of the form '$(\forall x)(Fx \,\&\, \sim Fx)$' which contains that quantifier initially would in this situation contain initially some quantifier that—as it happened—was equivalent to the explicitly defined numerical quantifier '$(\exists 100\alpha)(— \alpha —)$'; but clearly we do not want to have to conclude that such a proposition is also of the form '$(\exists n\alpha)(\Phi\alpha)$'. However, the criterion I have actually given is not subject to such an objection, just because the quantifiers in question are directly required to be pure quantifiers, and therefore it cannot be satisfied merely accidentally.

Even though the criterion given does indeed do the work intended of it, in so far as it does pick out just those propositions that we would want it to as of the form '$(\exists n\alpha)(\Phi\alpha)$', still perhaps one might feel that the addition of a necessity operator would somehow improve matters. But I think that this would in fact only tend to introduce obscurity. Certainly we could reasonably argue that the quantifier expressed in English by 'there are three hundred . . .' is *necessarily* equivalent to the corresponding explicitly defined numerical quantifier, for this equivalence could be demonstrated by analysis of the meaning of the expression 'three hundred'. On the other hand it simply is not clear what ground one could have for supposing that the *unanalysable* quantifier 'there are V . . .' is *necessarily* equivalent to the explicitly defined quantifier 'there are 5 . . .', except perhaps for one's conviction that if the two really are equivalent over the whole of their range then this could hardly be a mere 'coincidence'. But if it is the material equivalence that is the sole basis for any judgement of necessity here, then it seems altogether better to give the material equivalence as the criterion. (And if a necessary equivalence would be less suitable than a material one, then clearly a Leibnizian identity is out of the question.)

To return, then, to the original version of the criterion, as I have stated it so far it is that a proposition is of the form '$(\exists n\alpha)(\Phi\alpha)$' if and only if it is of some form '$(Q\alpha)(\Phi\alpha)$' where there is a true proposition of one of the forms

$$(\forall\phi)((Q\alpha)(\phi\alpha) \equiv (\exists 1\alpha)(\phi\alpha))$$
$$(\forall\phi)((Q\alpha)(\phi\alpha) \equiv (\exists 2\alpha)(\phi\alpha))$$
$$(\forall\phi)((Q\alpha)(\phi\alpha) \equiv (\exists 3\alpha)(\phi\alpha))$$

and so on. But it is now time to consider eliminating the intuitive 'and so on'.

First, let us formally introduce the relations of equivalence and succession between quantifiers by means of the definitions

'$(Q'\alpha)(\Phi\alpha)$' for '$(\exists\alpha)(\Phi\alpha \,\&\, (Q\beta)(\Phi\beta \,\&\, \beta \neq \alpha))$'

'$Q_1 \equiv Q_2$' for '$(\forall\phi)((Q_1\alpha)(\phi\alpha) \equiv (Q_2\alpha)(\phi\alpha))$'

'$Q_1 \mathbf{S} Q_2$' for '$(\forall\phi)((Q_1\alpha)(\phi\alpha) \equiv (Q_2'\alpha)(\phi\alpha))$'.

With the help of these definitions we may restate the condition that a quantifier, Q, should satisfy at least one of the theses

$$(\forall\phi)((Q\alpha)(\phi\alpha) \equiv (\exists 1\alpha)(\phi\alpha))$$
$$(\forall\phi)((Q\alpha)(\phi\alpha) \equiv (\exists 2\alpha)(\phi\alpha))$$
$$(\forall\phi)((Q\alpha)(\phi\alpha) \equiv (\exists 3\alpha)(\phi\alpha))$$

etc.

in a more succinct fashion as the condition that

$$Q \equiv \exists \vee Q \equiv \exists' \vee Q \equiv \exists'' \vee Q \equiv \exists''' \vee \ldots$$

But further, it is a simple matter to prove from our definitions that

$$Q_1 \equiv Q_2 \supset Q_1' \equiv Q_2'$$

while from our explanation of the schematic letter 'Q' it clearly follows that[12]

$$(\text{—}Q'\text{—}) \supset (\exists q)(\text{—}q\text{—})$$

and from these two facts it is easy to establish the equivalences

$$(Q \equiv \exists') \equiv Q\mathbf{S}\exists$$
$$(Q \equiv \exists'') \equiv (\exists q)(Q\mathbf{S}q \;\&\; q\mathbf{S}\exists)$$
$$(Q \equiv \exists''') \equiv (\exists q_1)(\exists q_2)(Q\mathbf{S}q_1 \;\&\; q_1\mathbf{S}q_2 \;\&\; q_2\mathbf{S}\exists)$$

etc.

Applying this transformation to the infinite disjunction

$$Q \equiv \exists' \vee Q \equiv \exists'' \vee Q \equiv \exists''' \vee \ldots$$

we at once see that it is simply the expansion of

$$Q{\star}\mathbf{S}\exists$$

where '... $\star\mathbf{S}$ ...' is the proper ancestral of '... $\mathbf{S}$ ...'. And if we agree to adopt the non-standard definition[13]

$$\text{'}Q_1{*}\mathbf{S}Q_2\text{'} \quad \text{for} \quad \text{'}Q_1 \equiv Q_2 \vee Q_1{\star}\mathbf{S}Q_2\text{'}$$

then we can represent the whole of our original condition as simply the condition that

$$Q{*}\mathbf{S}\exists.$$

So it turns out that our criterion may be put, without any explicit appeal to an 'and so on', as: a proposition is of the form '$(\exists n\alpha)(\Phi\alpha)$' if and only if there is some true proposition of the form '$Q{*}\mathbf{S}\exists$' which is such that our proposition is of the related form '$(Q\alpha)(\Phi\alpha)$'. But at this point we should turn aside to consider the notion, just introduced, of the ancestral of a relation.

3. The Ancestral

One of the most important characteristics of the ancestral of a relation is its connection with arguments by induction, which are constantly employed in reasoning about natural numbers. As

[12] See the last pages of chapter 4, section 4.

[13] The definition is non-standard because of course the improper ancestral is generally understood as though defined with identity in the first disjunct here, rather than equivalence.

I noted in chapter 1, the principle which legitimizes inductive arguments is (in its application to subject-variables)

$$\vdash (Fb \mathbin{\&} (\forall x)(\forall y)(xRy \supset (Fy \supset Fx))) \supset (\forall x)(x{*}Rb \supset Fx)$$

and an alternative way of writing this principle is

$$\vdash a{*}Rb \supset ((\forall x)(\forall y)(xRy \supset (Fy \supset Fx)) \supset (Fb \supset Fa)).$$

As I also noted in chapter 1, this principle follows from the definition of the ancestral, and indeed on the usual definition of the ancestral this is supremely obvious; for the definition usually given here is just

$$\text{‘}a{*}Rb\text{’} \quad \text{for} \quad \text{‘}(\forall \mathfrak{f})((\forall x)(\forall y)(xRy \supset (\mathfrak{f}y \supset \mathfrak{f}x)) \supset (\mathfrak{f}b \supset \mathfrak{f}a))\text{’}.$$

Roughly, then, we may say that the ancestral of a relation is simply *defined as* that function of the relation for which inductive arguments hold, and since natural numbers will always be defined in any logicist construction by means of the ancestral of the successor relation and the first number, say 1, it comes about that the natural numbers are simply defined as those entities for which inductive arguments based on 1 and the successor relation are valid. This, however, is a way of putting the matter which seems to bear no relation whatever to my discussion of the definition of numbers—or rather, of numerical quantifiers—in the last section. For there I was in effect taking the notion of the ancestral of a relation to be the notion of a relation which is given by an infinite disjunction. More precisely, I took it that the *proper* ancestral '... ⋆R ...' of the relation '... R ...' would be given by the infinite disjunction

$$a{\star}Rb \equiv aRb \vee (\exists x)(aRx \mathbin{\&} xRb) \vee (\exists x)(\exists y)(aRx \mathbin{\&} xRy \mathbin{\&} yRb) \vee \ldots$$

and I then proposed to define the improper ancestral '... *R ...' in terms of the proper ancestral. To clear up this matter, then, we ought to give some assurance that what is intuitively understood as an infinite disjunction is the same relation as would be formally defined quite differently by the definition connecting it with induction. As a matter of fact I shall propose a slightly different definition from that usually adopted, partly because it suits my procedure to consider the *proper* ancestral first,[14] and

[14] In fact the proper ancestral is clearly the more intuitive notion, as is reflected in the adoption of the relation '... is an ancestor of ...' as the eponymous paradigm.

partly because it seems worth pointing out that there is no necessity to rely explicitly on the inductive principle in constructing a definition. But the problem is still the same: it must be shown that the definition proposed does correspond to the infinite disjunction we intuitively understand.

It will save us from having to work with an unfamiliar type-neutral symbolism from the start if we begin by supposing that we have only *first-level* relations to consider, i.e. relations which hold between objects, and for these I use the schematic letters 'R', 'S', 'R_1',... with associated variables 'R', 'S', 'R_1',... . And it will save us from using excessively lengthy formulae if we employ relevant parts of the symbolism of the traditional logic of relations, which we may obtain by explicit definition

$$\text{`a(R}\vee\text{S)b'} \quad \text{for} \quad \text{`aRb} \vee \text{aSb'}$$

$$\text{`a(R|S)b'} \quad \text{for} \quad \text{`}(\exists x)(aRx \,\&\, xSb)\text{'}$$

$$\text{`R} \subset \text{S'} \quad \text{for} \quad \text{`}(\forall x)(\forall y)(xRy \supset xSy)\text{'}.$$

With this apparatus we can say that our intuitive starting point for the notion of the proper ancestral is that

$$a{*}Rb \equiv aRb \vee aR|Rb \vee aR|R|Rb \vee \ldots$$

or more simply that the relation '... *R ...' just is the relation '... (R ∨ R|R ∨ R|R|R ∨ ...)...', and our first step must be to establish some properties of the relation '... *R ...' by employing intuitive reasoning on the intuitively understood infinite disjunction.

For this purpose, let us first note that we can prove, by perfectly straightforward reasoning in quantification theory, each of the theorems on the infinite list

$$\vdash R|(S_1 \vee S_2) \subset (R|S_1 \vee R|S_2)$$

$$\vdash R|(S_1 \vee S_2 \vee S_3) \subset (R|S_1 \vee R|S_2 \vee R|S_3)$$

$$\vdash (R|(S_1 \vee S_2 \vee S_3 \vee S_4) \subset (R|S_1 \vee R|S_2 \vee R|S_3 \vee R|S_4)$$

etc.

and so generally we shall be able to prove any theorem of the form

$$\vdash R|(S_1 \vee S_2 \vee S_3 \vee \ldots) \subset (R|S_1 \vee R|S_2 \vee R|S_3 \vee \ldots)$$

where the disjunction '$S_1 \vee S_2 \vee S_3 \vee \ldots$' is of finite length. Now the intuitive step that we need to take is to see that the principle continues to hold even where the disjunction is infinite, though of course we cannot produce a formal proof in this case. And one might justify such a step in this way. If we say that anything x bears the relation '$R|(S_1 \vee S_2 \vee S_3 \ldots)$' to y, then this means that x bears the relation 'R' to something z which bears at least one of the relations 'S_1' or 'S_2' and 'S_3' or ... to y. Even if the list of relations 'S_1' or 'S_2' or 'S_3' or ... is infinite, nevertheless there will still be at least one of them, say 'S_n', which z bears to y, and therefore x will at least bear the relation '$R|S_n$' to y. But from this it follows immediately that x bears at least one of the relations '$R|S_1$' or '$R|S_2$' or ... '$R|S_n$' or ... to y, and hence that x bears the relation '$(R \mid S_1 \vee R \mid S_2 \vee R \mid S_3 \vee \ldots)$' to y. So I think we can see that the reasoning which supports this principle is quite unaffected by the infinity of the disjunction, and we may be sure that the principle holds in this case too, despite the lack of a formal proof in quantification theory.

Applying the principle to the infinite disjunction which we take as defining the proper ancestral, we may therefore assert

$$\vdash R|(R \vee R|R \vee R|R|R \vee \ldots) \subset (R|R \vee R|R|R \vee R|R|R|R \vee \ldots)$$

and from this it follows at once that

$$\vdash R \vee R|(R \vee R|R \vee R|R|R \ldots) \subset (R \vee R|R \vee R|R|R \vee R|R|R|R \vee \ldots)$$

or in other words $\vdash (R \vee R|{*R}) \subset {*R}$.

This reasoning establishes a property of the relation '$*R$' which we may also put by saying that '$*R$' satisfies the condition on 'S' that

$$(R \vee R|S) \subset S.$$

However, '$*R$' is not the only relation which satisfies this condition, so we might next consider what can be said of all relations 'S' that do satisfy it. Suppose, then, that 'S' is any relation which satisfies the condition, or—what comes to the same thing—suppose that 'S' satisfies the two conditions

$$R \subset S$$

$$R|S \subset S.$$

In that case it is easy to see that we may assert each of the propositions on the infinite list

$$R \subset S$$

$$R|R \subset S$$

$$R|R|R \subset S$$

$$R|R|R|R \subset S$$

etc.

For the first proposition on the list is simply the first of the conditions on 'S'; further, from the first proposition it immediately follows that

$$R|R \subset R|S$$

and from this the second proposition follows in view of the second condition on 'S'; then again the second proposition on the list at once implies

$$R|R|R \subset R|S$$

from which again the third proposition follows in view of the second condition on 'S'; and so on. Evidently each proposition on the list can be established from its predecessor in this way, and so we may be sure that they do all follow from our initial conditions on 'S'. But now from this position we can further establish the proposition

$$(R \vee R|R \vee R|R|R \vee \ldots) \subset S.$$

As before we can give a formal proof of this proposition from our initial assumptions in any case in which the disjunction is taken as finite, but we can also see that the point must still hold even where the disjunction is infinite. For our proposition states that for anything x and anything y, if x bears the relation '$(R \vee R|R \vee R|R|R \vee \ldots)$' to y then it also bears S to y. But if we say that x bears the relation '$(R \vee R|R \vee R|R|R \vee \ldots)$' to y, then this means that x bears at least one of the relations 'R' or 'R|R' or 'R|R|R' or ... to y, and whichever one it is we have already seen that we can prove that in that case x must also bear 'S' to y. So again the reasoning is quite unaffected by the infinity of the disjunction, and we may be sure that under the conditions proposed

$${}^{*}R \subset S.$$

I have now given intuitive arguments to establish the two theses concerning '*R':

A1 $\vdash: (R \vee R|{*R}) \subset {*R}$

A2 $\vdash: ((R \vee R|S) \subset S) \supset ({*R} \subset S)$.

We could obviously establish yet more theses in the same way, but in fact we have now employed our intuitions sufficiently, for all the properties of '*R' that we shall ever need to be concerned with are given by these two theses, and it would be entirely satisfactory to adopt just these as axioms for the notion. For it turns out that between them they determine the relation '*R' uniquely—or at least, to within an equivalence. Roughly speaking, we may say that the first thesis states that '*R' satisfies the condition on 'S' that '$(R \vee R|S) \subset S$', while the second states that '*R' is the *minimum* relation that satisfies this condition. More formally, we may observe that A1 and A2 between them imply the equivalence

$$\vdash a{*R}b \equiv (\forall s)(((R \subset s) \,\&\, (R|s \subset s)) \supset asb)$$

for the implication from left to right is easily seen to be a mere rewriting of A2 together with a step of generalisation licensed by quantification theory, while the implication from right to left is obtained by taking the universally quantified 's' in particular as '*R' and noting that A1 is then the antecedent to the conditional. It follows that any relation which satisfies the conditions stated in A1 and A2 will also satisfy this equivalence, and so be equivalent to '*R'. And it also follows that the intuitive reasoning that has established A1 and A2 has thereby established the truth of this equivalence, and so, we may suppose, it has established the adequacy of the definition

'a*Rb' for '$(\forall s)(((R \subset s) \,\&\, (R|s \subset s)) \supset asb)$'

and I shall adopt it forthwith.

At this point I should perhaps make it clear that I call the reasoning we have been employing 'intuitive' only because it is not reasoning that can be formally represented within orthodox quantification theory (except at a metalinguistic level), the reason being simply that that theory contains no means of handling an infinite disjunction. But I do not want to be taken as implying that the reasoning is in some way 'illogical', for at any rate it is no more 'illogical' than is the initial idea of an infinite disjunction; if we can make sense of that idea—as I am sure we can—then we must surely allow that it is contained in that idea that nothing can satisfy an infinite disjunction without satisfying at least one of the disjuncts, and it is almost entirely that principle that I have been relying on. But I postpone further discussion of this matter to the next chapter, and return now to the business in hand.

From the definition of the proper ancestral that I propose—or indeed from A1 and A2 taken as axioms—it is possible to deduce all the usual theorems for the proper ancestral by the methods of ordinary quantification theory,[15] and of course to deduce equivalences corresponding to such other definitions as have been proposed.[16] Further, since the definition is itself framed solely in terms of truthfunctional connectives and universal quantification, there is evidently no obstacle to rephrasing it in a type-neutral way so that it comes also to cover relations holding between quantifiers. In particular the proper ancestral of the successor relation amongst pure quantifiers will then be furnished with the definition

$$\text{`}Q_1{\star}\mathbf{S}Q_2\text{'} \quad \text{for} \quad \text{`}(\forall\mathfrak{R})(((\mathbf{S} \subset \mathfrak{R}) \,\&\, (\mathbf{S}|\mathfrak{R} \subset \mathfrak{R})) \supset Q_1\mathfrak{R}Q_2)\text{'}$$

where we have employed the abbreviations

$$\text{`}\mathbf{S} \subset \boldsymbol{\mathfrak{R}}\text{'} \quad \text{for} \quad \text{`}(\forall q_1)(\forall q_2)(q_1\mathbf{S}q_2 \supset q_1\boldsymbol{\mathfrak{R}}q_2)\text{'}$$

$$\text{`}\mathbf{S}|\boldsymbol{\mathfrak{R}} \subset \boldsymbol{\mathfrak{R}}\text{'} \quad \text{for} \quad \text{`}(\forall q_1)(\forall q_2)((\exists q_3)(q_1\mathbf{S}q_3 \,\&\, q_3\boldsymbol{\mathfrak{R}}q_2) \supset q_1\boldsymbol{\mathfrak{R}}q_2)\text{'}$$

and all the theorems developed for the proper ancestral of a relation holding between objects will transfer intact to the proper ancestral of the successor relation between quantifiers. As for the improper ancestral, I have already stated that in the case of the successor relation I adopt the non-standard definition

$$\text{`}Q_1{*}\mathbf{S}\,Q_2\text{'} \quad \text{for} \quad \text{`}Q_1 \equiv Q_2 \vee Q_1{\star}\mathbf{S}Q_2\text{'},$$

for this yields a conveniently simple way of summing up the criterion for containing a numerical quantifier. Since the definition is a non-standard one, I regard it as laid down only for the successor relation and in no other case, but it does anyway

[15] The theorems I have had occasion to use in my construction of arithmetic are furnished with proofs in Appendix 1 to this chapter.

[16] For instance Frege's definition, in his *Foundations of Arithmetic*, §§ 79–80, of 'y follows in the ϕ-series after x'.

suffice for all those theorems involving the improper ancestral that I have had occasion to use.[17]

The non-standard nature of the definition does, however, alert us to a complication which arises over the inductive principle. Among the theorems that we establish for the proper ancestral of the successor relation will be[18]

$$\vdash (\Phi Q \,\&\, (\forall q_1)(\forall q_2)(q_1 \mathbf{S} q_2 \supset (\Phi q_2 \supset \Phi q_1))) \supset (\forall q)(q{\star}\mathbf{S}Q \supset \Phi q)$$

and for the purposes of induction we shall need to make use of the special case of this theorem

$$\vdash (\Phi \exists \,\&\, (\forall q_1)(\forall q_2)(q_1 \mathbf{S} q_2 \supset (\Phi q_2 \supset \Phi q_1))) \supset (\forall q)(q{\star}\mathbf{S}\exists \supset \Phi q).$$

Now if we make the assumption that the context 'Φ ...' which we are dealing with is an extensional context in the sense that

$$(\forall q_1)(\forall q_2)(q_1 \equiv q_2 \supset (\Phi q_1 \equiv \Phi q_2))$$

we shall obviously be able to assert in addition that

$$\Phi \exists \supset (\forall q)(q \equiv \exists \supset \Phi q)$$

while from our definitions it will be a simple matter to show that

$$(\forall q)(\Phi q \supset \Phi q') \supset (\forall q_1)(\forall q_2)(q_1 \mathbf{S} q_2 \supset (\Phi q_2 \supset \Phi q_1)).$$

Consequently on this assumption our original theorem will imply

$$(\Phi \exists \,\&\, (\forall q)(\Phi q \supset \Phi q')) \supset (\forall q)(q \equiv \exists \supset \Phi q) \,\&\, (\forall q)(q{\star}\mathbf{S}\exists \supset \Phi q)$$

and therefore, in view of the definition of the improper ancestral,

$$(\Phi \exists \,\&\, (\forall q)(\Phi q \supset \Phi q')) \supset (\forall q)(q{*}\mathbf{S}\exists \supset \Phi q).$$

It is essentially this principle which legitimizes the use of inductive arguments, but it is to be noted that the principle holds only on the condition that 'Φ ...' is extensional. I do not regard this condition as unwelcome. On the contrary my discussion of numerical identity in the next section will make it clear that such a condition is entirely to be expected, and of course it will always be satisfied wherever we do wish to make use of an inductive argument. For the moment, then, let us just note the condition and continue.

For I should observe, before closing this section, that inductive arguments concerning numbers will not take quite the form of the principle just mentioned. To obtain the principle in the form in

[17] Theorems A7–A9. [18] Theorem A4.

which it will actually be used we must first substitute for the schematic letter 'Φ' so as to obtain

$$(\Phi\exists \,\&\, \exists{*}\mathbf{S}\exists \,\&\, (\forall q)(\Phi q \,\&\, q{*}\mathbf{S}\exists \supset \Phi q' \,\&\, q'{*}\mathbf{S}\exists)) \supset (\forall q)(q{*}\mathbf{S}\exists \supset \Phi q \,\&\, q{*}\mathbf{S}\exists)$$

which we may then simplify by noting that we are already in possession of the theorems

$$\vdash \exists{*}\mathbf{S}\exists$$

$$\vdash Q{*}\mathbf{S}\exists \supset Q'{*}\mathbf{S}\exists$$

since the first is an immediate consequence of '$\exists \equiv \exists$' and the second follows at once from '$Q'SQ$'. So finally we reach the version

$$(\Phi\exists \,\&\, (\forall q)(q{*}\mathbf{S}\exists \supset (\Phi q \supset \Phi q'))) \supset (\forall q)(q{*}\mathbf{S}\exists \supset \Phi q).$$

This version evidently entitles us to make use of arguments of the form.

If	$\vdash (\text{—}\exists\text{—})$
and	$\vdash Q{*}\mathbf{S}\exists \supset ((\text{—}Q\text{—}) \supset (\text{—}Q'\text{—}))$
then	$\vdash Q{*}\mathbf{S}\exists \supset (\text{—}Q\text{—})$

(wherever the context '(—...—)' is extensional), and these arguments may be abbreviated to

if	$\vdash (\text{—}1\text{—})$
and	$\vdash (\text{—}n\text{—}) \supset (\text{—}n'\text{—})$
then	$\vdash (\text{—}n\text{—})$

which is how inductive arguments are usually written. But this has evidently brought us on to a new topic, viz. the statement of theorems involving numerical letters and variables.

4. Arithmetic

In section 2 we reached the position that a proposition should be said to be of the form '$(\exists n\alpha)(\Phi\alpha)$' if and only if it is of some form '$(Q\alpha)(\Phi\alpha)$' which is such that the proposition of the related form '$Q{*}\mathbf{S}\exists$' is true. In this way the elementary numerical form '$(\exists n\alpha)(\Phi\alpha)$' is introduced as a special case of the form '$(Q\alpha)(\Phi\alpha)$', and in just the same way we may introduce more complicated numerical forms as special cases of the correspondingly complicated forms written with pure quantifier letters. In fact quite generally a proposition will be of any numerical form '(... (∃nα)(—α—)...)' if and only if it is of a corresponding form '(...(Qα)(—α—)...)' for which the proposition of the related form '$Q*\mathbf{S}\exists$' is true. And on the basis of this criterion we may develop our formal treatment of numerical quantifiers by giving an explicit definition of all those formulae involving them that we shall actually need to consider.

For if we have some formula written with a numerical schematic letter, say

$$(...(\exists n\alpha)(—\alpha—)...),$$

then we have just said that to assert the truth of all propositions of that form is the same as to assert the truth of all propositions of the corresponding form

$$(...(Q\alpha)(—\alpha—)...)$$

which are such that the proposition of the related form '$Q*\mathbf{S}\exists$' is true, and that is to say that it is to assert the truth of all propositions of the corresponding form

$$Q*\mathbf{S}\exists \supset (...(Q\alpha)(—\alpha—)...).$$

More concisely, what I have just said is that any two propositions of the corresponding forms

$$\vdash (...(\exists n\alpha)(—\alpha—)...)$$

$$\vdash (Q*\mathbf{S}\exists \supset (...(Q\alpha)(—\alpha—)...))$$

are interchangeable, and we may therefore *define* the first formula as an abbreviation for the second. This gives us a definition of numerical schematic letters only in contexts of a certain special sort, namely those beginning with '⊢', but in our *formal* development all the contexts in which such schematic letters occur will be of this special sort.

The same treatment can evidently be extended to numerical variables. A variable of any sort will always occur bound to a quantifier, and the first case to consider is where the quantifier is a universal quantifier. Here evidently we lay down the analogous definition

$$\text{'}(\forall \mathfrak{n})(...(\exists \mathfrak{n}\alpha)(—\alpha—)...)\text{'}$$

for

$$\text{'}(\forall q)(q*\mathbf{S}\exists \supset (...(q\alpha)(—\alpha—)...))\text{'}.$$

With these two definitions it is easily seen that the laws of ordinary quantification theory—Generalization, Instantiation, and Distribution; also alphabetic change of letter or variable—may be applied to our numerical letters and variables just as if they were quite ordinary letters and variables. Further, supposing the existential quantifier to have been already defined in terms of the universal quantifier for ordinary variables, it will be appropriate to lay down for numerical variables the definition

$$\text{`}(\exists\mathfrak{n})(\ldots(\exists\mathfrak{n}\alpha)(-\alpha-)\ldots)\text{' for `}(\exists q)(q{*}\mathbf{S}\exists\ \&\ (\ldots(q\alpha)(-\alpha-)\ldots))\text{'}$$

whereupon we shall evidently deduce

$$\vdash(\exists\mathfrak{n})(\ldots(\exists\mathfrak{n}\alpha)(-\alpha-)\ldots)\ \equiv\ \sim(\forall\mathfrak{n})(\sim(\ldots(\exists\mathfrak{n}\alpha)(-\alpha-)\ldots))$$

so that the usual relations hold here too. There is one further use of numerical variables that we shall require, for in the proof of Frege's theorem these variables occur bound not to the universal quantifier or the existential quantifier but to further numerical quantifiers. The appropriate definition to give here will evidently be one with the general effect that

$$(\exists n\mathfrak{m})(-\mathfrak{m}-)$$

is defined as an abbreviation for

$$(\exists nq)(q{*}\mathbf{S}\exists\ \&\ (-q-))$$

but before we frame this more exactly we should pause to remove the limitation, which I imposed at the end of section 1, to numerical quantifiers based on non-identity as their counting relation.

For in the construction of arithmetic which I sketched in my first chapter, after the initial step of introducing numerical letters and variables and showing that the ordinary rules of quantification theory applied to them, the next step was to lay down definitions for the notion of being the same number and being a successor of a number. For the first of these I originally suggested a definition which I wrote as

$$\text{`}n{=}m\text{' for `}(\forall\mathfrak{f})((\exists nx)(\mathfrak{f}x)\equiv(\exists mx)(\mathfrak{f}x))\text{'}$$

and at a later stage I considered the possibility of exchanging this definition for

$$\text{`}n{=}m\text{' for `}(\forall\phi)((\exists n\mathfrak{p})(\phi\mathfrak{p})\equiv(\exists m\mathfrak{p})(\phi\mathfrak{p}))\text{'.}$$

But now we can see that it is obviously arbitrary to define numerical identity by confining our attention only to *subject*-quantifiers, or by confining it to *number*-quantifiers, and so the definiens we require should evidently be written in a type-neutral way. And we can now also see that it is almost equally arbitrary to restrict our attention to this or that special counting relation, and so it seems clear that the definiens we require here is in fact

$$(\forall\mathfrak{R})(\forall\phi)((\exists n_{\mathfrak{R}}\,\alpha)(\phi\alpha) \equiv (\exists m_{\mathfrak{R}}\,\alpha)(\phi\alpha)).$$

The identity defined by this definiens is evidently not to be identified with the Leibnizian identity we defined in section 1, and so should not be symbolized by the same sign '... = ...'. Nor is this identity a case of the equivalence between quantifiers that we defined in section 2, for the quantifiers in this formula are represented by the expressions '$\exists n_{\mathfrak{R}}$' and '$\exists m_{\mathfrak{R}}$', and since these are not expressions for pure quantifiers—for the relation '$\mathfrak{R}$' may well be a type-restricted one—that definition of equivalence is strictly not applicable here. But it is convenient to extend that definition to certain impure quantifiers by adding the further clause

$$\text{'}Q_{1\mathfrak{R}} \equiv Q_{2\mathfrak{R}}\text{'} \quad \text{for} \quad \text{'}(\forall\phi)((Q_{1\mathfrak{R}}\,\alpha)(\phi\alpha) \equiv (Q_{2\mathfrak{R}}\,\alpha)(\phi\alpha))\text{'}$$

and then, introducing the new sign '$\ldots =_N \ldots$' to represent the relation of being the same number as, we may define

$$\text{'}n =_N m\text{'} \quad \text{for} \quad \text{'}(\forall\mathfrak{R})(\exists n_{\mathfrak{R}} \equiv \exists m_{\mathfrak{R}}).$$

Although the relation '$\ldots =_N \ldots$' as I here define it is not itself a case of the relation of equivalence between quantifiers, it is in a certain sense equivalent to the relation of equivalence between certain quantifiers. For, to reflect the fundamental role which Leibnizian identity has to play in our concept of a numerical quantifier, we defined numerical quantifiers in general by means of the definition

$$\text{'}(\exists n_{\mathfrak{R}}\,\alpha)(\text{—}\alpha\text{—})\text{'} \quad \text{for} \quad \text{'}(\exists\psi)((\exists n\alpha)(\psi\alpha) \;\&\; (\forall\alpha)(\forall\beta)(\psi\alpha \;\&\; \psi\beta \;\&\; \alpha \neq \beta \supset \alpha\mathfrak{R}\beta) \;\&\; (\forall\alpha)(\psi\alpha \supset \text{—}\alpha\text{—}))\text{'}$$

and from this it is easy to see that we may now assert[19]

$$\vdash n =_N m \equiv (\exists n_{\neq} \equiv \exists m_{\neq}).$$

Indeed it is only because this equivalence holds that we could properly confine our attention to numerical quantifiers based on non-identity in the previous sections, for the reason why it is sufficient to require that a proposition contain a quantifier *equivalent* to some quantifier '$(\exists n_{\neq}\alpha)(\text{—}\alpha\text{—})$' if it is itself to contain a quantifier '$(\exists n_{\neq}\,\alpha)(\text{—}\alpha\text{—})$' is just that equivalence

[19] See Appendix 2, Theorem 0.6, for a formal proof.

between *these* numerical quantifiers is sufficient to ensure sameness of number. But non-identity is not the only counting relation that has this property. For instance, the arguments adduced in section 2 would apply equally well to numerical quantifiers based on non-equivalence, and it is not hard to show that

$$\vdash n =_N m \equiv (\exists n_{\not\equiv} \equiv \exists m_{\not\equiv}).$$

Again, Frege's theorem (which we shall shortly be establishing) effectively constitutes a proof that

$$\vdash n =_N m \equiv (\exists n_{\neq_N} \equiv \exists m_{\neq_N})$$

and examples could easily be multiplied. The relation '$\ldots =_N \ldots$' as I define it simply remains neutral between these various equivalences between certain numerical quantifiers, but its extra generality does not represent any added strength.

It will be seen that numerical identity, as here defined, cannot be guaranteed to license substitution in all contexts without exception, but only in those contexts that are built up by extensional methods (such as truth-functional combination and universal quantification) from contexts of the form

$$\text{`}(\exists n_{\mathfrak{R}}\, \alpha)(\text{—}\, \alpha\, \text{—})\text{'}.$$

Without an axiom for the purpose—which I do not provide—we cannot assert

$$\vdash n =_N m \supset (\Phi n \equiv \Phi m)$$

and it seems to me that this is all to the good. After all, it is not implausible to hold that there are propositions concerning, say, 7+5 which differ in truth-value from the corresponding propositions concerning 12, and there seems no reason to believe that we are ruling out this possibility when we assert that 7+5 *is equal to* 12, which is all that we mean by calling them *the same number*. (And it is for this reason that I said in the last section that the restriction on inductive arguments which I there mentioned was altogether to be expected, for of course inductive arguments issue in conclusions about *all* numbers—about 7+5 no less than 12.) What we do expect to be licensed by a numerical equality, or identity, is at least substitution in all *arithmetical* contexts, and this much we do certainly have from our definition, since all arithmetical contexts are extensional in the required way.

So far, then, we have offered definitions of certain formulae containing numerical quantifiers based on non-identity, namely those of the forms[20]

$$\vdash (\ldots(\exists n\alpha)(-\alpha-)\ldots)$$
$$(\forall \mathfrak{n})(\ldots(\exists \mathfrak{n}\alpha)(-\alpha-)\ldots)$$
$$(\exists \mathfrak{n})(\ldots(\exists \mathfrak{n}\alpha)(-\alpha-)\ldots).$$

We have also defined other numerical quantifiers in terms of the numerical quantifiers based on non-identity, and we shall now set down the definitions

$$\text{'}n =_N m\text{'} \quad \text{for} \quad \text{'}(\forall \mathfrak{R})(\exists n_{\mathfrak{R}} \equiv \exists m_{\mathfrak{R}})\text{'}$$
$$\text{'}n\mathbf{S}_N m\text{'} \quad \text{for} \quad \text{'}(\forall \mathfrak{R})(\exists n_{\mathfrak{R}} \equiv \exists m'_{\mathfrak{R}})\text{'}.$$

For the purposes of formal proof, however, we shall mainly make use of the derived equivalences[21]

$$\vdash n =_N m \equiv (\exists n \equiv \exists m)$$
$$\vdash n\mathbf{S}_N m \equiv (\exists n\ \mathbf{S}\ \exists m).$$

We shall also require to make use of the ancestrals of the new successor relation '... $\mathbf{S}_N$ ...'. Its improper ancestral may evidently be defined in terms of its proper ancestral

$$\text{'}n{*}\mathbf{S}_N m\text{'} \quad \text{for} \quad \text{'}n =_N m \vee n{\star}\mathbf{S}_N m\text{'}$$

and for the proper ancestral it seems most appropriate to adopt a definition in which our variables are limited throughout to numerical variables, so that the definition has the force of the infinite disjunction

$$\vdash n{\star}\mathbf{S}_N m \equiv n\mathbf{S}_N m \vee (\exists \mathfrak{p})(n\mathbf{S}_N \mathfrak{p}\ \&\ \mathfrak{p}\mathbf{S}_N m) \vee (\exists \mathfrak{p}_1)(\exists \mathfrak{p}_2)(\ldots.$$

This will allow us to use the ancestral relations in our arithmetical proofs without at any point needing to consider variables other than numerical variables, but it will still preserve the useful equivalence between statements about numbers proper and statements about numerical quantifiers based on non-identity. For if we apply our definition of '$\vdash(-n-)$' to the simple theorem[22]

$$\vdash Q_1{*}\mathbf{S}\exists \supset (Q\mathbf{S}Q_1 \supset Q{*}\mathbf{S}\exists)$$

we obtain

$$\vdash Q\mathbf{S}\exists m \supset Q{*}\mathbf{S}\exists$$

[20] To avoid circularity at this point, the dummy variable 'α' must not be understood to be replaceable by *numerical* variables in these definition-schemata, but only by variables already fully introduced. See further pp. 175–6 below.

[21] Appendix 2, Theorems 0.6, and 0.7.

[22] Theorem A8.

and so, considering just the second disjunct of our expansion, it will follow that

$$\vdash (\exists q)(q{*}S\exists \,\&\, \exists nSq \,\&\, qS\exists m) \equiv (\exists q)(\exists nSq \,\&\, qS\exists m)$$

whence, by our definition of '$(\exists \mathfrak{n})(\text{—}\,\mathfrak{n}\,\text{—})$',

$$\vdash (\exists \mathfrak{p})(\exists nS\exists \mathfrak{p} \,\&\, \exists \mathfrak{p}S\exists m) \equiv (\exists q)(\exists nSq \,\&\, qS\exists m)$$

whence finally, by the equivalences just noted,

$$\vdash (\exists \mathfrak{p})(nS_N\mathfrak{p} \,\&\, \mathfrak{p}S_Nm) \equiv (\exists q)(\exists nSq \,\&\, qS\exists m).$$

Applying this argument to each of the disjuncts in our expansion in turn, we can see intuitively that a similar equivalence will hold for each of them, and therefore it will be the case that

$$\begin{aligned} \vdash n{\star}S_Nm &\equiv nS_Nm \vee (\exists \mathfrak{p})(nS_N\mathfrak{p} \,\&\, \mathfrak{p}S_Nm) \vee (\exists \mathfrak{p}_1)(\exists \mathfrak{p}_2)(\ldots \\ &\equiv \exists n\ S\ \exists m \vee (\exists q)(\exists nSq \,\&\, qS\exists m) \vee (\exists q_1)(\exists q_2)(\ldots \\ &\equiv \exists n\ {\star}S\ \exists m. \end{aligned}$$

To achieve this result formally we have only to define the proper ancestral '$\ldots {\star}S_N \ldots$' by taking the standard definition of '$\ldots {\star}S \ldots$' with its quantifiers restricted throughout to numerical variables, and so one's first inclination would be to set down

'$n{\star}S_Nm$' for $(\forall \mathfrak{R})(((S_N \subset \mathfrak{R}) \,\&\, (S_N|\mathfrak{R} \subset \mathfrak{R})) \supset n\mathfrak{R}m)$

where we employ the abbreviations

'$S_N \subset \mathfrak{R}$' for '$(\forall \mathfrak{n})(\forall \mathfrak{m})(\mathfrak{n}S_N\mathfrak{m} \supset \mathfrak{n}\mathfrak{R}\mathfrak{m})$'

'$S_N|\mathfrak{R} \subset \mathfrak{R}$' for '$(\forall \mathfrak{n})(\forall \mathfrak{m})((\exists \mathfrak{p})(\mathfrak{n}S_N\mathfrak{p} \,\&\, \mathfrak{p}\mathfrak{R}\mathfrak{m}) \supset \mathfrak{n}\mathfrak{R}\mathfrak{m})$'.

But I think that a doubt might arise about this formulation since the formula '$n\mathfrak{R}m$' might appear to be undefined, as it does not restrict the numerical letters or variables to contexts for which we have already defined them. However, since we are in fact resolved to define all contexts for 'n' in terms of the primary context '$\exists n$' it will evidently be no loss of generality if we rewrite the clause '$n\mathfrak{R}m$' everywhere in this definition as '$\exists n\mathfrak{R}\exists m$', and I shall accordingly adopt the definition with this modification. It will then be possible to produce a formal proof of the desired equivalence[23]

$$\vdash n{\star}S_Nm \equiv \exists n\, {\star}S\, \exists m$$

whence also $\vdash n{*}S_Nm \equiv \exists n\, {*}S\, \exists m$

[23] Appendix 2, Theorem 0.8.

This completes the definitions that we shall require for the first stage of our construction of arithmetic, and it is now time to consider the axioms. The original axioms should now of course be rewritten in the form

$$\vdash (\exists \mathfrak{n})(\mathfrak{n} =_N 1)$$
$$\vdash (\exists \mathfrak{n})(\mathfrak{n} S_N m)$$
$$\vdash n {*} S_N 1.$$

Applying the equivalences just noted, these three theses reduce to

$$\vdash (\exists \mathfrak{n})(\exists \mathfrak{n} \equiv \exists 1)$$
$$\vdash (\exists \mathfrak{n})(\exists \mathfrak{n} \equiv \exists m')$$
$$\vdash \exists n \ {*}S \ \exists 1$$

and applying the definitions of numerical letters and variables. and of '∃1', these in turn become

$$\vdash (\exists q)(q {*} S \exists \ \& \ q \equiv \exists)$$
$$\vdash Q {*} S \exists \supset (\exists q)(q {*} S \exists \ \& \ q \equiv Q')$$
$$\vdash Q {*} S \exists \supset Q {*} S \exists$$

which are very easily established as theorems. The last has indeed become a quite trivial tautology, while in view of the definition of the ancestral we may clearly assert[24]

$$\vdash \exists {*} S \exists$$
$$\vdash Q_1 {*} S \exists \supset (Q_2 S Q_1 \supset Q_2 {*} S \exists)$$

and in view of the definitions of '... S ...' and 'Q'' we may also assert

$$\vdash Q_1' S Q_1$$

but these are all the premisses we need to establish

$$\vdash \exists {*} S \exists \ \& \ \exists \equiv \exists$$
$$\vdash Q {*} S \exists \supset (Q' {*} S \exists \ \& \ Q' \equiv Q')$$

from which our axioms follow by existential generalization. As regards this existential generalization (and the step above whereby 'Q_1'' was substituted for 'Q_2') the principles involved here are simply

$$\vdash (-\exists-) \supset (\exists q)(-q-)$$
$$\vdash (-Q'-) \supset (\exists q)(-q-)$$

24 Theorems A7 and A8.

and these are obvious consequences of the substitution rules noticed in chapter 4, section 4.

Under the new definitions, then, the axioms of our original construction come to be truths of logic, and the same evidently holds of the first six theorems. For it is now an entirely trivial matter to rewrite the original proofs to conform to the new definitions, and—with the help of the inductive principle established in the last section—there is no difficulty in seeing that the proofs remain valid under this rewriting.

Turning now to the proof of Frege's theorem, our object here is to count numbers, and for this purpose we should evidently employ the counting relation '... $\neq_N$ ...' occurring in formulae of the pattern

$$(\exists n_{\neq_N} \mathfrak{m})(\text{—}\,\mathfrak{m}\,\text{—}).$$

It should be observed that such formulae are not already furnished with a definition. For although we have laid down a definition for all formulae of the pattern

$$(\exists n_{\mathfrak{R}}\alpha)(\text{—}\,\alpha\,\text{—})$$

this definition itself involves the corresponding formula

$$(\exists n\alpha)(\text{—}\,\alpha\,\text{—})$$

which is still not defined for a numerical variable in place of 'α'. Our previous definitions inform us that

$$\vdash (\ldots(\exists n\alpha)(\text{—}\,\alpha\,\text{—})\ldots)$$

is in general an abbreviation for

$$\vdash (Q{*}\mathbf{S}\exists \supset (\ldots(Q\alpha)(\text{—}\,\alpha\,\text{—})\ldots))$$

and if this were taken as covering numerical variables in place of 'α' we would be supposing that we already understood the definiens

$$\vdash (Q{*}\mathbf{S}\exists \supset (\ldots(Q\mathfrak{n})(\text{—}\,\mathfrak{n}\,\text{—})\ldots)).$$

But we do not. The part '$(Q\mathfrak{n})(\text{—}\,\mathfrak{n}\,\text{—})$' has not so far been defined, nor is it easy to see how to frame the general definition that would be required, for we have already seen that its two special cases '$(\forall\mathfrak{n})(\text{—}\,\mathfrak{n}\,\text{—})$' and '$(\exists\mathfrak{n})(\text{—}\,\mathfrak{n}\,\text{—})$' have to be given divergent definitions.

It must be understood, then, that the previous definitions apply only to formulae containing a part '$(\exists n\alpha)(\text{—}\,\alpha\,\text{—})$' where 'α' stands in place of a genuine variable that is already fully introduced independently of our new and contextually defined numerical 'variables'. Consequently we now require a fresh definition for formulae containing numerical variables in this position, where the counting relation is taken as the relation of not being the same number as. The condition of adequacy for the definition is evidently that it shall yield the recursive equivalences

$$\vdash (\exists 1_{\neq_N}\mathfrak{m})(-\mathfrak{m}-) \equiv (\exists \mathfrak{m})(-\mathfrak{m}-)$$

$$\vdash (\exists n'_{\neq_N}\mathfrak{m})(-\mathfrak{m}-) \equiv (\exists \mathfrak{m})(-\mathfrak{m}- \,\&\, (\exists n_{\neq_N}\mathfrak{p})(-\mathfrak{p}- \,\&\, \mathfrak{p} \neq_N \mathfrak{m}).$$

In view of the primacy of the context '$\exists\mathfrak{m}$' we may limit our attention here to the cases

$$\vdash (\exists 1_{\neq_N}\mathfrak{m})(\Phi(\exists\mathfrak{m})) \equiv (\exists\mathfrak{m})(\Phi(\exists\mathfrak{m}))$$

$$\vdash (\exists n'_{\neq_N}\mathfrak{m})(\Phi(\exists\mathfrak{m})) \equiv (\exists\mathfrak{m})(\Phi(\exists\mathfrak{m}) \,\&\, (\exists n_{\neq_N}\mathfrak{p})(\Phi(\exists\mathfrak{p}) \,\&\, \mathfrak{p}_{\neq_N}\mathfrak{m})).$$

And these reduce by our previous definitions to

$$\vdash (\exists 1_{\neq_N}\mathfrak{m})(\Phi(\exists\mathfrak{m})) \equiv (\exists q)(q{*}\mathbf{S}\exists \,\&\, \Phi q)$$

$$\vdash (\exists n'_{\neq_N}\mathfrak{m})(\Phi(\exists\mathfrak{m}))$$

$$\equiv (\exists q)(q{*}\mathbf{S}\exists \,\&\, \Phi q \,\&\, (\exists n_{\neq_N}\mathfrak{p})(\Phi(\exists\mathfrak{p}) \,\&\, \exists\mathfrak{p} \neq q)).$$

It is easily seen that these last two equivalences are consequences of the definition

'$(\exists n_{\neq_N}\mathfrak{m})(\Phi(\exists\mathfrak{m}))$' for '$(\exists n_{\neq} q)(q{*}\mathbf{S}\exists \,\&\, \Phi q)$'

which I therefore adopt.

The actual proof of Frege's theorem now offers no difficulty. To see first how the two lemmata of the original proof function with the new definitions, let us consider the first application of the first lemma (in theorem 7, line 5) where it is invoked to justify the theorem we now write as

$$\vdash (\forall\mathfrak{m})(n{*}\mathbf{S}_N\mathfrak{m} \supset n'{*}\mathbf{S}_N\mathfrak{m})$$

$$\supset ((\exists n_{\neq_N}\mathfrak{m})(n{*}\mathbf{S}_N\mathfrak{m}) \supset (\exists n_{\neq_N}\mathfrak{m})(n'{*}\mathbf{S}_N\mathfrak{m})).$$

Here we note that the lemma we are now dealing with has been established in the more general form

$$\vdash (\forall\alpha)(\Phi\alpha \supset \Psi\alpha) \supset ((\exists n_{\mathfrak{R}}\,\alpha)(\Phi\alpha) \supset (\exists n_{\mathfrak{R}}\,\alpha)(\Psi\alpha))$$

and we proceed by first taking the special case of this lemma in which the variables are quantifier-variables and the counting relation non-equivalence, so as to obtain

$$\vdash (\forall q)(\Phi q \supset \Psi q) \supset ((\exists n_{\neq}\,q)(\Phi q) \supset (\exists n_{\neq}\,q)(\Psi q))$$

and by next substituting for ‘Φ...’ and ‘Ψ...’ to get

$$\vdash (\forall q)((\exists n * \mathbf{S} q \,\&\, q * \mathbf{S} \exists) \supset (\exists n' * \mathbf{S} q \,\&\, q * \mathbf{S} \exists)) \supset ((\exists n_{\neq} q)(\exists n * \mathbf{S} q \,\&\, q * \mathbf{S} \exists) \supset (\exists n_{\neq} q)(\exists n' * \mathbf{S} q \,\&\, q * \mathbf{S} \exists)).$$

Applying a simple transformation to the antecedent of this theorem we deduce

$$\vdash (\forall q)(q * \mathbf{S} \exists \supset (\exists n * \mathbf{S} q \supset \exists n' * \mathbf{S} q)) \supset ((\exists n_{\neq} q)(\exists n * \mathbf{S} q \,\&\, q * \mathbf{S} \exists) \supset (\exists n_{\neq} q)(\exists n' * \mathbf{S} q \,\&\, q * \mathbf{S} \exists))$$

and this abbreviates by our definitions to

$$\vdash (\forall \mathfrak{m})(\exists n * \mathbf{S} \exists \mathfrak{m} \supset \exists n' * \mathbf{S} \exists \mathfrak{m}) \supset ((\exists n_{\neq_N} \mathfrak{m})(\exists n * \mathbf{S} \exists \mathfrak{m}) \supset (\exists n_{\neq_N} \mathfrak{m})(\exists n' * \mathbf{S} \exists \mathfrak{m})$$

which finally yields the theorem desired in view of the equivalence between ‘$\exists n * \mathbf{S} \exists m$’ and ‘$n * \mathbf{S}_N m$’.[25]

The remaining applications of the first lemma are entirely similar, and so too is the application of the second lemma (which, though it will not be forthcoming for *all* counting relations, is easily shown to hold for the counting relation of non-equivalence between quantifiers). One further point which I earlier singled out for comment was the use (in theorem 7 line 7, and in theorem 9 line 3) of the thesis

$$\vdash n \geqslant_N n' \supset ((\exists n_{\neq_N} \mathfrak{m})(n * \mathbf{S}_N \mathfrak{m}) \supset (\exists n'_{\neq_N} \mathfrak{m})(n * \mathbf{S}_N \mathfrak{m}))$$

but it turns out that the justification for this is again exactly similar. For the antecedant is now an abbreviation for

$$(\forall \mathfrak{R})(\forall \phi)((\exists n_{\mathfrak{R}} \alpha)(\phi \alpha) \supset (\exists n'_{\mathfrak{R}} \alpha)(\phi \alpha))$$

and this antecedent clearly implies

$$(\forall \phi)((\exists n_{\neq} q)(\phi q) \supset (\exists n'_{\neq} q)(\phi q))$$

which in turn implies

$$(\exists n_{\neq} q)(\exists n * \mathbf{S} q \,\&\, q * \mathbf{S} \exists) \supset (\exists n'_{\neq} q)(\exists n * \mathbf{S} q \,\&\, q * \mathbf{S} \exists)$$

[25] Actually, to avoid begging the question in this last step, we should have taken as our substitution for ‘Φq’ not a mere equivalent of ‘$n * \mathbf{S}_N \mathfrak{m}$’ but its full definition in this context, namely

$$q * \mathbf{S} \exists \,\&\, ((\forall \mathfrak{R})(\exists n_{\mathfrak{R}} \equiv q_{\mathfrak{R}}) \vee (\forall \mathfrak{R})(((\forall \mathfrak{n})(\forall \mathfrak{m})(\mathfrak{n} \mathbf{S}_N \mathfrak{m} \supset \exists \mathfrak{n} \,\mathfrak{R}\, \exists \mathfrak{m}) \,\&\, (\forall \mathfrak{n})(\forall \mathfrak{m})(\forall \mathfrak{p})(\mathfrak{n} \mathbf{S}_N \mathfrak{p} \,\&\, \exists \mathfrak{p} \,\mathfrak{R}\, \exists \mathfrak{m} \supset \exists \mathfrak{n} \,\mathfrak{R}\, \exists \mathfrak{m})) \supset \exists n \,\mathfrak{R} q))$$

See Appendix 2, Theorem 0.9, with the preceding discussion.

which abbreviates to

$$(\exists n_{\neq_N}\mathfrak{m})(\exists n *\mathbf{S}\ \exists\mathfrak{m}) \supset (\exists n'_{\neq_N}\mathfrak{m})(\exists n *\mathbf{S}\ \exists\mathfrak{m})$$

and from this again the desired theorem follows in view of the same equivalence.

There are now no more stumbling blocks in the deduction of Frege's theorem, nor therefore in the establishing of Peano's postulates. With our better understanding of the notion of a numerical quantifier, and with our consequently revised definitions of the arithmetical formulae we require, we do now find that the theses which constitute the foundation of arithmetic have been transformed by our definitions into purely logical truths. And so it is that Frege's original programme of deducing arithmetic from logic, without any additional axioms, has to all appearance been successfully completed.

Appendix 1. Theorems for the Ancestral

In this appendix I employ parts of the traditional symbolism of the logic of classes and relations, which I obtain by the abbreviations[26]

'a(R∨S)b'	for	'(aRb ∨ aSb)'
'a(R↾F)b'	for	'(aRb & Fb)'
'a(F↿R)b'	for	'(Fa & aRb)'
'a(R|S)b'	for	'(∃x)(aRx & xSb)'
'(R''F)a'	for	'(∃x)(aRx & Fx)'
'F ⊂ G'	for	'(∀x)(Fx ⊃ Gx)'
'F = G'	for	'(∀x)(Fx ≡ Gx)'
'R ⊂ S'	for	'(∀x)(∀y)(xRy ⊃ xSy)'
'R = S'	for	'(∀x)(∀y)(xRy ≡ xSy)'

The proper ancestral is of course defined by the definition

D✲: 'a✲Rb' for ('∀s)(((R ⊂ s) & (R|s ⊂ s)) ⊃ asb)'

Apart from the deduction of theorems A1 and A2, which make obvious use of quantification over relations, all the transitions in the proofs that follow depend solely on first-order quantification theory.

[26] The use of '... = ...' here defined should be distinguished from the Leibnizian identity defined in section 1 of this chapter, and I should strictly introduce a different symbol for this role. However, there is no danger of confusion, for the present usage of '... = ...' is confined to this appendix, and this appendix never employs Leibnizian identity.

(i)

I begin by recovering the fundamental theorems A1 and A2 from the definition, and adding a simple corollary.

THEOREM A1(a). $R \subset *R$

1.	— aRb	Hyp
2.	— — $(R \subset S)\ \&\ (R|S \subset S)$	Hyp
3.	— — $R \subset S$	2
4.	— — $aRb \supset aSb$	3
5.	— — aSb	1, 4
6.	— $((R \subset S)\ \&\ (R|S \subset S)) \supset aSb$	2–5
7.	— $(\forall s)(((R \subset s)\ \&\ (R|s \subset s)) \supset asb)$	1–6
8.	— $a{*}Rb$	7, D$*$

THEOREM A1(b). $R|*R \subset *R$

1.	— aRb	Hyp
2.	— — $(R \subset S)\ \&\ (R|S \subset S)$	Hyp
3.	— — $R|S \subset S$	2
4.	— — $(aRb\ \&\ bSc) \supset aSc$	3
5.	— — $bSc \supset aSc$	1, 4
6.	— $((R \subset S)\ \&\ (R|S \subset S)) \supset (bSc \supset aSc)$	2–5
7.	— $(((R \subset S)\ \&\ (R|S \subset S)) \supset bSc) \supset (((R \subset S)\ \&\ (R|S \subset S)) \supset aSc)$	6
8.	— $(\forall s)(((R \subset s)\ \&\ (R|s \subset s)) \supset bsc)$ $\supset (\forall s)(((R \subset s)\ \&\ (R|s \subset s)) \supset asc)$	1–7
9.	— $b{*}Rc \supset a{*}Rc$	8, D$*$

THEOREM A2. $((R \vee R|S) \subset S) \supset (*R \subset S)$

1.	$a{*}Rb \supset (((R \subset S)\ \&\ (R|S \subset S)) \supset aSb)$	D$*$
2.	$((R \subset S)\ \&\ (R|S \subset S)) \supset (a{*}Rb \supset aSb)$	1
3.	$((R \subset S)\ \&\ (R|S \subset S)) \supset (*R \subset S)$	2

THEOREM A3. $*R = (R \vee R|*R)$

1.	$(R \vee R|*R) \subset *R$	A 1
2.	$R|(R \vee R|*R) \subset R|*R$	1
3.	$(R \vee R|(R \vee R|*R)) \subset (R \vee R|*R)$	2
4.	$*R \subset (R \vee R|*R)$	3, A2

I now proceed to the main principle of induction. This is given in a general form in A4, but I add afterwards two special cases that prove convenient.

THEOREM A4. $(R``F \subset F) \supset (*R``F \subset F)$

1.	$(R \subset S) \supset (R|*S \subset S|*S)$	—
2.	$\supset ((R \vee R|*S) \subset (S \vee S|*S))$	1
3.	$\supset ((R \vee R|*S) \subset *S$	2, A 1
4.	$\supset (*R \subset *S)$	3, A 2

5.	$(S \mid S \subset S) \supset ((S \vee S \mid S) \subset S)$	—
6.	$\supset ({\ast}S \subset S)$	5, A 2
7.	$\supset ((R \subset S) \supset ({\ast}R \subset S))$	4, 6
8.	$(\forall x)(\forall y)((\exists z)((Fz \supset Fx) \& (Fy \supset Fz)) \supset (Fy \supset Fx))$	—
9.	$(\forall x)(\forall y)(xRy \supset (Fy \supset Fx)) \supset (\forall x)(\forall y)(x{\ast}Ry \supset (Fy \supset Fx))$	7, 8
10.	$(\forall x)((\exists y)(xRy \& Fy) \supset Fx) \supset (\forall x)((\exists y)(x{\ast}Ry \& Fy) \supset Fx)$	9,

THEOREM A4(a). $(R \mid S \subset S) \supset ({\ast}R \mid S \subset S)$

1.	$(\forall x)(\forall y)(xRy \& Fy \supset Fx) \supset (\forall x)(\forall y)(x{\ast}Ry \& Fy \supset Fx)$	A 4
2.	$(\forall x)(\forall y)(xRy \& ySa \supset xSa) \supset (\forall x)(\forall y)(x{\ast}Ry \& ySa \supset xSa)$	1, sub
3.	$(\forall x)(\forall y)(\forall z)(xRy \& ySz \supset xSz)$ $\supset (\forall x)(\forall y)(\forall z)(x{\ast}Ry \& ySz \supset xSz)$	2

THEOREM A4(b). $(S \mid R \subset S) \supset (S \mid {\ast}R \subset S)$

1.	$(\forall x)(\forall y)(\forall z)(xRy \& \sim zSy \supset \sim zSx)$ $\supset (\forall x)(\forall y)(\forall z)(x{\ast}Ry \& \sim zSy \supset \sim zSx)$	A 4(a)
2.	$(\forall x)(\forall y)(\forall z)(zSx \& xRy \supset zSy)$ $\supset (\forall x)(\forall y)(\forall z)(zSx \& x{\ast}Ry \supset zSy)$	1

The only further theorem on the proper ancestral that was actually used in the construction of arithmetic was '$R \mid {\ast}R = {\ast}R \mid R$'. This is given below as A6, being preceded by two lemmas which are stronger forms of the induction principle

THEOREM A5(a). $(R \mid S \subset S) \supset ({\ast}R \mid S \subset R \mid S)$

1.	$(R \mid S \subset S) \supset ({\ast}R \mid S \subset S)$	A 4 (a)
2.	$\supset (R \mid {\ast}R \mid S \subset R \mid S)$	1
3.	$\supset ((R \mid S \vee R \mid {\ast}R \mid S) \subset R \mid S)$	2
4.	$\supset ((R \vee R \mid {\ast}R) \mid S \subset R \mid S)$	3
5.	$\supset ({\ast}R \mid S \subset R \mid S)$	4, A 3

THEOREM A5(b). $(S \mid R \subset S) \supset (S \mid {\ast}R \subset S \mid R)$

1.	$(S \mid R \subset S) \supset (S \mid R \mid R \subset S \mid R)$	—
2.	$\supset (S \mid R \mid {\ast}R \subset S \mid R)$	1, A 4 (b)
3.	$\supset ((S \mid R \vee S \mid R \mid {\ast}R) \subset S \mid R)$	2
4.	$\supset (S \mid (R \vee R \mid {\ast}R) \subset S \mid R)$	3
5.	$\supset (S \mid {\ast}R \subset S \mid R)$	4, A 3

THEOREM A6. $R \mid {\ast}R = {\ast}R \mid {\ast}R = {\ast}R \mid R$

1.	$R \subset {\ast}R$	A 1 (a)
2.	$R \mid {\ast}R \subset {\ast}R \mid {\ast}R$	1
3.	${\ast}R \mid R \subset {\ast}R \mid {\ast}R$	1
4.	$R \mid {\ast}R \subset {\ast}R$	A 1, (b)
5.	${\ast}R \mid {\ast}R \subset R \mid {\ast}R$	4, A 5 (a)
6.	${\ast}R \mid R \subset {\ast}R$	3, 5, 4
7.	${\ast}R \mid {\ast}R \subset {\ast}R \mid R$	6, A 5 (b)
8.	$R \mid {\ast}R = {\ast}R \mid {\ast}R = {\ast}R \mid R$	2, 5; 3, 7

(ii)

A general treatment of the improper ancestral of a relation is complicated by the fact that I adopt two non-standard definitions of the improper ancestrals '... $\star \mathbf{S}$ ...' and '... $\star \mathbf{S_N}$ ...'. I shall therefore proceed here in a schematic way, supposing that the improper ancestral is defined in terms of some relation '... I ...' which I do not further specify. The definition takes the form

D*: '$\ast R$' for '$(I \vee \star R)$'

and I impose upon '... I ...' the condition

C(I): $I|R = R = R|I$

(This condition is of course satisfied for the two definitions of an improper ancestral that I actually employ, and I think it would be satisfied for any plausible definitions of an improper ancestral.)

The three theorems for the improper ancestral that I actually employ may now be very simply deduced

Theorem A7. $\ast R = (I \vee \star R)$

Immediate from D*

Theorem A8. $(R \vee R|\ast R) \subset \ast R$

1.	$(R \vee R\|\ast R) \subset (R \vee R\|(I \vee \star R))$	D*
2.	$\subset (R \vee R\|I \vee R\|\star R)$	1
3.	$\subset (R \vee R \vee R\|\star R)$	2, C(I)
4.	$\subset (R \vee R\|\star R)$	3
5.	$\subset \star R$	4, A 1
6.	$\subset \ast R$	5, D*

Theorem A9. $R|\ast R = \ast R|R$

1.	$R\|\ast R = R\|(I \vee \star R)$	D*
2.	$= R\|I \vee R\|\star R$	1
3.	$= I\|R \vee R\|\star R$	2, C(I)
4.	$= I\|R \vee \star R\|R$	3, A 6
5.	$= (I \vee \star R)\|R$	4
6.	$= \ast R\|R$	5, D*

(iii)

Finally, in preparation for the deductions of the next appendix, I establish a theorem, A11, which shows how in certain conditions the original definition of the proper ancestral can be exchanged for a more restricted one. A10 is a lemma for A11.

Theorem A10. $(R``F \subset F) \supset ((\ast R)\upharpoonright F \subset \ast(R\upharpoonright F))$

1. — $R``F \subset F$ Hyp
2. — $R\upharpoonright F = F\upharpoonleft R\upharpoonright F$ 1
3. — $(R\upharpoonright F \vee (R\upharpoonright F)\mid\ast(R\upharpoonright F)) = (F\upharpoonleft R\upharpoonright F \vee (F\upharpoonleft R\upharpoonright F)\mid\ast(R\upharpoonright F))$ 2
4. — $(R\upharpoonright F \vee (R\upharpoonright F)\mid\ast(R\upharpoonright F)) = F\upharpoonleft(R\upharpoonright F \vee (R\upharpoonright F)\mid\ast(R\upharpoonright F))$ 3
5. — $\ast(R\upharpoonright F) = F\upharpoonleft\ast(R\upharpoonright F)$ 4, A3
6. — $R\mid\ast(R\upharpoonright F) = R\mid(F\upharpoonleft\ast(R\upharpoonright F))$ 5
7. — $R\mid\ast(R\upharpoonright F) = (R\upharpoonright F)\mid\ast(R\upharpoonright F)$ 6
8. — $(R\upharpoonright F \vee R\mid\ast(R\upharpoonright F)) = (R\upharpoonright F \vee (R\upharpoonright F)\mid\ast(R\upharpoonright F))$ 7
9. — $(R\upharpoonright F \vee R\mid\ast(R\upharpoonright F)) = \ast(R\upharpoonright F)$ 8, A3
10. — $((R\mid I)\upharpoonright F \vee R\mid\ast(R\upharpoonright F)) = \ast(R\upharpoonright F)$ 9, C(I)
11. — $(R\mid(I\upharpoonright F) \vee R\mid\ast(R\upharpoonright F)) = \ast(R\upharpoonright F)$ 10
12. — $R\mid(I\upharpoonright F \vee \ast(R\upharpoonright F)) = \ast(R\upharpoonright F)$ 11
13. — $R\mid(I\upharpoonright F \vee \ast(R\upharpoonright F)) \subset (I\upharpoonright F \vee \ast(R\upharpoonright F))$ 12
14. — $\ast R\mid(I\upharpoonright F \vee \ast(R\upharpoonright F)) \subset R\mid(I\upharpoonright F \vee \ast(R\upharpoonright F))$ 13, A5(a)
15. — $\ast R\mid(I\upharpoonright F \vee \ast(R\upharpoonright F)) \subset \ast(R\upharpoonright F)$ 14, 12
16. — $(\ast R\mid(I\upharpoonright F) \vee \ast R\mid\ast(R\upharpoonright F)) \subset \ast(R\upharpoonright F)$ 15
17. — $\ast R\mid(I\upharpoonright F) \subset \ast(R\upharpoonright F)$ 16
18. — $(\ast R\mid I)\upharpoonright F \subset \ast(R\upharpoonright F)$ 17
19. — $(\ast R)\upharpoonright F \subset \ast(R\upharpoonright F)$ 18, C(I)

Theorem A11. $(R``F \subset F) \supset (Fb \supset$
$(a\ast Rb \equiv (\forall s)(((F\upharpoonleft R\upharpoonright F \subset s) \,\&\, ((F\upharpoonleft R\upharpoonright F)\mid(s\upharpoonright F) \subset s)) \supset asb)))$

1. $R \subset \ast R$ A1
2. $R\upharpoonright F \subset \ast R$ 1
3. $\ast R\upharpoonright F \subset \ast R$ —
4. $(R\upharpoonright F)\mid(\ast R\upharpoonright F) \subset \ast R\mid\ast R$ 2, 3
5. $(R\upharpoonright F)\mid(\ast R\upharpoonright F) \subset \ast R$ 4, A6, A1
6. $(R\upharpoonright F \subset \ast R) \,\&\, ((R\upharpoonright F)\mid(\ast R\upharpoonright F) \subset \ast R)$ 2, 5
7. $(\forall s)(((R\upharpoonright F \subset s) \,\&\, ((R\upharpoonright F)\mid(s\upharpoonright F) \subset s)) \supset asb) \supset a\ast Rb$ 6
8. — $R``F \subset F$ Hyp
9. — $R\upharpoonright F = F\upharpoonleft R\upharpoonright F$ 8
10. — — Fb Hyp
11. — — — $a\ast Rb$ Hyp
12. — — — $a((\ast R)\upharpoonright F)b$ 10, 11
13. — — — — $(R\upharpoonright F \subset S) \,\&\, ((R\upharpoonright F)\mid(S\upharpoonright F) \subset S)$ Hyp
14. — — — — $(R\upharpoonright F \subset S\upharpoonright F) \,\&\, ((R\upharpoonright F)\mid(S\upharpoonright F) \subset S\upharpoonright F)$ 13
15. — — — — $\ast(R\upharpoonright F) \subset S\upharpoonright F$ 14, A2
16. — — — — $\ast(R\upharpoonright F) \subset S$ 15
17. — — — — $(\ast R)\upharpoonright F \subset S$ 8,16, A10
18. — — — — aSb 12, 17
19. — — — $((R\upharpoonright F \subset S) \,\&\, ((R\upharpoonright F)\mid(S\upharpoonright F) \subset S)) \supset aSb$ 13–18
20. — — — $(\forall s)(((R\upharpoonright F \subset s) \,\&\, ((R\upharpoonright F))\mid(s\upharpoonright F) \subset s)) \supset asb)$ 11–19
21. — — $a\ast Rb \supset (\forall s)(((R\upharpoonright F \subset s) \,\&\, ((R\upharpoonright F)\mid(s\upharpoonright F) \subset s)) \supset asb)$ 11–20
22. — — $a\ast Rb \equiv (\forall s)(((R\upharpoonright F \subset s) \,\&\, ((R\upharpoonright F)\mid(s\upharpoonright F) \subset s)) \supset asb)$ 7, 21
23. — — $(a\ast Rb \equiv (\forall s)(((F\upharpoonleft R\upharpoonright F \subset s) \,\&\,$
$((F\upharpoonleft R\upharpoonright F)\mid(s\upharpoonright F) \subset s)) \supset asb))$ 9, 21

Appendix 2. Auxiliary Theorems for Arithmetic

The purpose of this appendix is to give formal proofs of the theorems involving numerical letters and variables which were cited without proof in the body of this chapter. I begin by recalling, and labelling, the relevant definitions.

First, Leibnizian identity is defined, for all propositional components, by the definition

D(=): $\quad$ 'A=B' for '$(\forall\phi)(\phi A \equiv \phi B)$'.

Employing this definition we then introduce a particular successor relation between pure quantifiers, along with the relation of equivalence between these quantifiers, by the definitions

D(Q′): $\quad$ '$(Q'\alpha)(\Phi\alpha)$' for '$(\exists\alpha)(\Phi\alpha \,\&\, (Q\beta)(\Phi\beta \,\&\, \beta\neq\alpha))$'

D(≡): $\quad$ '$Q_1 \equiv Q_2$' for '$(\forall\phi)((Q_1\alpha)(\phi\alpha) \equiv (Q_2\alpha)(\phi\alpha))$'

D(S): $\quad$ '$Q_1\mathbf{S}Q_2$' for '$(\forall\phi)((Q_1\alpha)(\phi\alpha) \equiv (Q_2'\alpha)(\phi\alpha))$'.

The proper ancestral '... ***S** ...' of this relation '... **S** ...' is defined in the standard way for the proper ancestral of a relation between quantifiers, while the improper ancestral is defined

$$\text{'}Q_1{*}\mathbf{S}Q_2\text{' for '}Q_1 \equiv Q_2 \vee Q_1{}^{*}\mathbf{S}Q_2\text{'}.$$

It is then a trivial matter to show that the condition C(I) of the previous appendix is satisfied in the form

$$\vdash (\exists q)(Q_1\mathbf{S}q \,\&\, q\equiv Q_2) \equiv Q_1\mathbf{S}Q_2 \equiv (\exists q)(Q_1\equiv q \,\&\, q\mathbf{S}Q_2)$$

and hence that the theorems there developed for the improper ancestral hold for the relation '... ***S** ...' just as the theorems developed for the proper ancestral hold for '... ***S** ...'. In particular, in view of the discussion of inductive arguments in section 3 of this chapter, it is easy to see how the proof of Lemma 1 of the appendix to chapter 1 may be adapted to yield a proof of the thesis

$$\vdash Q{*}\mathbf{S}\exists \supset ((\forall\alpha)(\Phi\alpha \supset \Psi\alpha) \supset ((Q\alpha)(\Phi\alpha) \supset (Q\alpha)(\Psi\alpha)))$$

and I shall employ this thesis, again under the title 'Lemma 1', in the proofs that follow.[27]

Turning now to formulae containing numerical letters and variables, we have first the contextual definitions

D(n): $\quad$ '$\vdash(...(\exists n\alpha)(-\alpha-)...)$' for '$\vdash(Q{*}\mathbf{S}\exists \supset (...(Q\alpha)(-\alpha-)...))$'

D($\mathfrak{n}$): $\quad$ '$(\forall\mathfrak{n})(...(\exists\mathfrak{n}\alpha)(-\alpha-)...)$' for '$(\forall q)(q{*}\mathbf{S}\exists \supset (...(q\alpha)(-\alpha-)...))$'

[27] Actually the thesis will mostly be used in the form

$$\vdash Q{*}\mathbf{S}\exists \supset ((Q\alpha)(\Phi\alpha) \equiv (\exists\psi)((Q\alpha)(\psi\alpha) \,\&\, (\forall\alpha)(\psi\alpha \supset \Phi\alpha)))$$

which is an entirely trivial reformulation of the original.

which permit us to employ the numerical quantifier expressions '$\exists$n' and '$\exists\mathfrak{n}$' in contexts already available for quantifier letters and variables, and next the definition

D($Q_{\mathfrak{R}}$): '$(Q_{\mathfrak{R}}\,\alpha)(-\alpha-)$' for

$$`(\exists\psi)((Q\alpha)(\psi\alpha)\ \&\ (\forall\alpha\beta)(\psi\alpha\ \&\ \psi\beta\ \&\ \alpha\neq\beta\supset\alpha\mathfrak{R}\beta)\ \&\ (\forall\alpha)(\psi\alpha\supset-\alpha-))'$$

which extends this permission more generally to the schematic expressions '$\exists n_{\mathfrak{R}}$' and '$\exists n'_{\mathfrak{R}}$', and their associated variables. I should also mention at this point two auxiliary definitions

D(1): '$(\exists 1\,\alpha)(-\alpha-)$' for '$(\exists\alpha)(-\alpha-)$'

D(count): 'Count($\mathfrak{R}$)' for '$(\forall\alpha)(\sim\alpha\mathfrak{R}\alpha)\ \&\ (\forall\alpha\beta)(\alpha\mathfrak{R}\beta\equiv\beta\mathfrak{R}\alpha)$'

and a further clause to be definition of equivalence

D($\equiv$): '$Q_1\mathfrak{R}\equiv Q_2\mathfrak{R}$' for '$(\forall\phi)((Q_{1\mathfrak{R}}\alpha)(\phi\alpha)\equiv(Q_{2\mathfrak{R}}\alpha)(\phi\alpha))$'.

The definition D($Q_{\mathfrak{R}}$) just given allows us to form the numerical quantifiers '$\exists n_{\neq}$' and '$\exists n'_{\neq}$' where the counting relation is explicitly given as non-identity, and the first point to be proved is that these are simply the original quantifiers '$\exists$n' and '$\exists$n′' under another name. The deduction here is very simple:

THEOREM 0·1. $\exists n\equiv\exists n_{\neq}$

1. $Q*\mathbf{S}\exists\supset((Q\alpha)(\Phi\alpha)\equiv(\exists\psi)((Q\alpha)(\psi\alpha)\ \&\ (\forall\alpha)(\psi\alpha\supset\Phi\alpha)))$ — Lemma 1
2. $(\exists n\alpha)(\Phi\alpha)\equiv(\exists\psi)((\exists n\alpha)(\psi\alpha)\ \&\ (\forall\alpha)(\psi\alpha\supset\Phi\alpha))$ — 1, D(n)
3. $(\exists n\alpha)(\Phi\alpha)\equiv(\exists\psi)((\exists n\alpha)(\psi\alpha)\ \&\ (\forall\alpha\beta)(\psi\alpha\ \&\ \psi\beta\ \&\ \alpha\neq\beta\supset\alpha\neq\beta)\ \&\ (\forall\alpha)(\psi\alpha\supset\Phi\alpha))$ — 2
4. $(\exists n\alpha)(\Phi\alpha)\equiv(\exists n_{\neq}\alpha)(\Phi\alpha)$ — 3, D($Q_{\mathfrak{R}}$)
5. $\exists n\equiv\exists n_{\neq}$ — 4, D($\equiv$)

THEOREM 0·2. $\exists n'\equiv\exists n'_{\neq}$

1. $Q'\mathbf{S}Q$ — D(Q′), D(**S**)
2. $Q*\mathbf{S}\exists\supset Q'*\mathbf{S}\exists$ — 1, A8
3. $Q*\mathbf{S}\exists\supset((Q'\alpha)(\Phi\alpha)\equiv(\exists\psi)((Q'\alpha)(\psi\alpha)\ \&\ (\forall\alpha)(\psi\alpha\supset\Phi\alpha)))$ — 2, Lemma 1
4. $(\exists n'\alpha)(\Phi\alpha)\equiv(\exists\psi)((\exists n'\alpha)(\psi\alpha)\ \&\ (\forall\alpha)(\psi\alpha\supset\Phi\alpha))$ — 3, D(n)
5. $(\exists n'\alpha)(\Phi\alpha)\equiv(\exists\psi)((\exists n'\alpha)(\psi\alpha)\ \&\ (\forall\alpha\beta)(\psi\alpha\ \&\ \psi\beta\ \&\ \alpha\neq\beta\supset\alpha\neq\beta)\ \&\ (\forall\alpha)(\psi\alpha\supset\Phi\alpha))$ — 4
6. $(\exists n'\alpha)(\Phi\alpha)\equiv(\exists n'_{\neq}\alpha)(\Phi\alpha)$ — 5, D($Q_{\mathfrak{R}}$)
7. $\exists n'\equiv\exists n'_{\neq}$ — 6, D($\equiv$)

The next point to be proved is that the definitions of '$\exists n_{\mathfrak{R}}$' and '$\exists n'_{\mathfrak{R}}$' are adequate, in the sense that they do yield the appropriate recursive equivalences (at least under the condition 'Count($\mathfrak{R}$)'). The first of these is very easily derived.

THEOREM 0·3. $(\exists 1_{\mathfrak{R}}\alpha)(\Phi\alpha) \equiv (\exists\alpha)(\Phi\alpha)$

1.	$— (\exists 1_{\mathfrak{R}}\alpha)(\Phi\alpha)$	Hyp
2.	$— (\exists\psi)((\exists 1\alpha)(\psi\alpha) \,\&\, (\forall\alpha\beta)(\psi\alpha \,\&\, \psi\beta \,\&\, \alpha\neq\beta \supset \alpha\mathfrak{R}\beta) \,\&\, (\forall\alpha)(\psi\alpha \supset \Phi\alpha))$	1, D($Q_{\mathfrak{R}}$)
3.	$— (\exists\psi)((\exists\alpha)(\psi\alpha) \,\&\, (\forall\alpha\beta)(\psi\alpha \,\&\, \psi\beta \,\&\, \alpha\neq\beta \supset \alpha\mathfrak{R}\beta) \,\&\, (\forall\alpha)(\psi\alpha \supset \Phi\alpha))$	2, D(1)
4.	$— (\exists\psi)((\exists\alpha)(\psi\alpha) \,\&\, (\forall\alpha)(\psi\alpha \supset \Phi\alpha))$	3
5.	$— (\exists\alpha)(\Phi\alpha)$	4
6.	$(\exists 1_{\mathfrak{R}}\alpha)(\Phi\alpha) \supset (\exists\alpha)(\Phi\alpha)$	1–5
7.	$— (\exists\alpha)(\Phi\alpha)$	Hyp
8.	$— (\exists\alpha)((\exists\beta)(\beta=\alpha) \,\&\, (\forall\beta\gamma)(\beta=\alpha \,\&\, \gamma=\alpha \,\&\, \beta\neq\gamma \supset \beta\mathfrak{R}\gamma) \,\&\, (\forall\beta)(\beta=\alpha \supset \Phi\beta))$	7, D(=)
9.	$— (\exists\alpha)((\exists 1\beta)(\beta=\alpha) \,\&\, (\forall\beta\gamma)(\beta=\alpha \,\&\, \gamma=\alpha \,\&\, \beta\neq\gamma \supset \beta\mathfrak{R}\gamma) \,\&\, (\forall\beta)(\beta=\alpha \supset \Phi\beta))$	8, D(1)
10.	$— (\exists\psi)((\exists 1\beta)(\psi\beta) \,\&\, (\forall\beta\gamma)(\psi\beta \,\&\, \psi\gamma \,\&\, \beta\neq\gamma \supset \beta\mathfrak{R}\gamma) \,\&\, (\forall\beta)(\psi\beta \supset \Phi\beta))$	9
11.	$— (\exists 1_{\mathfrak{R}}\,\alpha)(\Phi\alpha)$	10, D($Q_{\mathfrak{R}}$)
12.	$(\exists\alpha)(\Phi\alpha) \supset (\exists 1_{\mathfrak{R}}\,\alpha)(\Phi\alpha)$	7–11
13.	$(\exists\alpha)(\Phi\alpha) \equiv (\exists 1_{\mathfrak{R}}\,\alpha)(\Phi\alpha)$	6, 12

The second is more complex, and I take it in two parts:[28]

THEOREM 0·4. $(\exists n'_{\mathfrak{R}}\,\alpha)(\Phi\alpha) \supset (\exists\alpha)(\Phi\alpha \,\&\, (\exists n_{\mathfrak{R}}\,\beta)(\Phi\beta \,\&\, \beta\mathfrak{R}\alpha))$

1.	$— (\exists n'_{\mathfrak{R}}\,\alpha)(\Phi\alpha)$	Hyp
2.	$— (\exists\psi)((\exists n'\alpha)(\psi\alpha) \,\&\, (\forall\alpha\beta)(\psi\alpha \,\&\, \psi\beta \,\&\, \alpha\neq\beta \supset \alpha\mathfrak{R}\beta) \,\&\, (\forall\alpha)(\psi\alpha \supset \Phi\alpha)$	1, D($Q_{\mathfrak{R}}$)
3.	$— (\exists\psi)(\exists\alpha)(\psi\alpha \,\&\, (\exists n\beta)(\psi\beta \,\&\, \beta\neq\alpha) \,\&\, (\forall\alpha\beta)(\psi\alpha \,\&\, \psi\beta \,\&\, \alpha\neq\beta \supset \alpha\mathfrak{R}\beta) \,\&\, (\forall\alpha)(\psi\alpha \supset \Phi\alpha))$	2, D(Q')
4.	$— — \Psi A \,\&\, (\exists n\beta)(\Psi\beta \,\&\, \beta\neq A) \,\&\, (\forall\alpha\beta)(\Psi\alpha \,\&\, \Psi\beta \,\&\, \alpha\neq\beta \supset \alpha\mathfrak{R}\beta) \,\&\, (\forall\alpha)(\Psi\alpha \supset \Phi\alpha)$	Hyp
5.	$— — \Psi A \,\&\, (\exists n\beta)(\Psi\beta \,\&\, \beta\neq A) \,\&\, (\forall\beta)(\Psi\beta \,\&\, \Psi A \,\&\, \beta\neq A \supset \beta\mathfrak{R}A)$	4
6.	$— — (\exists n\beta)(\Psi\beta \,\&\, \beta\neq A) \,\&\, (\forall\beta)(\Psi\beta \,\&\, \beta\neq A \supset \beta\mathfrak{R}A)$	5
7.	$— — (\exists n\beta)(\Psi\beta \,\&\, \beta\neq A) \,\&\, (\forall\beta)(\Psi\beta \,\&\, \beta\neq A \supset \Psi\beta \,\&\, \beta\mathfrak{R}A)$	6
8.	$— — (\exists n\beta)(\Psi\beta \,\&\, \beta\mathfrak{R}A)$	7, Lemma 1
9.	$— — (\forall\alpha\beta)(\Psi\alpha \,\&\, \Psi\beta \,\&\, \alpha\neq\beta \supset \alpha\mathfrak{R}\beta)$	4
10.	$— — (\forall\alpha\beta)((\Psi\alpha \,\&\, \alpha\mathfrak{R}A) \,\&\, (\Psi\beta \,\&\, \beta\mathfrak{R}A) \,\&\, \alpha\neq\beta \supset \alpha\mathfrak{R}\beta)$	9
11.	$— — (\forall\beta)(\Psi\beta \supset \Phi\beta)$	4
12.	$— — (\forall\beta)((\Psi\beta \,\&\, \beta\mathfrak{R}A) \supset (\Phi\beta \,\&\, \beta\mathfrak{R}A))$	11
13.	$— — (\exists\psi)((\exists n\beta)(\psi\beta) \,\&\, (\forall\alpha\beta)(\psi\alpha \,\&\, \psi\beta \,\&\, \alpha\neq\beta \supset \alpha\mathfrak{R}\beta) \,\&\, (\forall\beta)(\psi\beta \supset \Phi\beta \,\&\, \beta\mathfrak{R}A)$	8, 10, 12
14.	$— — (\exists n_{\mathfrak{R}}\,\beta)(\Phi\beta \,\&\, \beta\mathfrak{R}A)$	13, D($Q_{\mathfrak{R}}$)

[28] I have simplified the presentation of this and the following proofs by using the methods of natural deduction, which involves using numerical schematic letters in contexts for which they are—strictly speaking—not defined. However, I regard this manner of presenting proofs merely as a convenient short cut, and the anomalous uses would of course disappear if the proofs were transformed by the standard methods to axiomatic proofs.

15.	— — $\Psi A \,\&\, (\forall\alpha)(\Psi\alpha \supset \Phi\alpha)$	4
16.	— — ΦA	15
17.	— — $\Phi A \,\&\, (\exists n_{\mathfrak{R}}\,\beta)(\Phi\beta \,\&\, \beta\mathfrak{R}A)$	14, 16
18.	— — $(\exists\alpha)(\Phi\alpha \,\&\, (\exists n_{\mathfrak{R}}\beta)(\Phi\beta \,\&\, \beta\mathfrak{R}\alpha))$	17
19.	— $(\exists\alpha)(\Phi\alpha \,\&\, (\exists n_{\mathfrak{R}}\beta)(\Phi\beta \,\&\, \beta\mathfrak{R}\alpha))$	3, 4–18
20.	$(\exists n'_{\mathfrak{R}}\,\alpha)(\Phi\alpha) \supset (\exists\alpha)(\Phi\alpha \,\&\, (\exists n_{\mathfrak{R}}\,\beta)(\Phi\beta \,\&\, \beta\mathfrak{R}\alpha))$	1–19

THEOREM 0·5. $\mathrm{Count}(\mathfrak{R}) \supset ((\exists\alpha)(\Phi\alpha \,\&\, (\exists n_{\mathfrak{R}}\,\beta)(\Phi\beta \,\&\, \beta\mathfrak{R}\alpha)) \supset (\exists n'_{\mathfrak{R}}\,\alpha)(\Phi\alpha))$

1.	— Count $(\mathfrak{R})$	Hyp
2.	— $(\forall\alpha)(\sim\alpha\mathfrak{R}\alpha) \,\&\, (\forall\alpha\beta)(\alpha\mathfrak{R}\beta \equiv \beta\mathfrak{R}\alpha)$	1, D (count)
3.	— — $(\exists\alpha)(\Phi\alpha \,\&\, (\exists n_{\mathfrak{R}}\,\beta)(\Phi\beta \,\&\, \beta\mathfrak{R}\alpha))$	Hyp
4.	— — $(\exists\alpha)(\exists\psi)(\Phi\alpha \,\&\, (\exists n\beta)(\psi\beta) \,\&\, (\forall\alpha\beta)(\Psi\alpha \,\&\, \Psi\beta \,\&\, \alpha\neq\beta \supset \alpha\mathfrak{R}\beta) \,\&\, (\forall\beta)(\psi\beta \supset \Phi\beta \,\&\, \beta\mathfrak{R}\alpha))$	3, D($Q_{\mathfrak{R}}$)
5.	— — — $\Phi A \,\&\, (\exists n\beta)(\Psi\beta) \,\&\, (\forall\alpha\beta)(\Psi a \,\&\, \Psi\beta \,\&\, \alpha\neq\beta \supset \alpha\mathfrak{R}\beta) \,\&\, (\forall\beta)(\Psi\beta \supset \Phi\beta \,\&\, \beta\mathfrak{R}A)$	Hyp
6.	— — — $(\exists n\beta)(\Psi\beta) \,\&\, (\forall\beta)(\Psi\beta \supset \Phi\beta \,\&\, \beta\mathfrak{R}A) \,\&\, (\forall\alpha)(\sim\alpha\mathfrak{R}\alpha)$	5, 2
7.	— — — $(\exists n\beta)(\Psi\beta) \,\&\, (\Psi A \supset \Phi A \,\&\, A\mathfrak{R}A) \,\&\, \sim A\mathfrak{R}A$	6
8.	— — — $(\exists n\beta)(\Psi\beta) \,\&\, \sim\Psi A$	7
9.	— — — $(\exists n\beta)(\Psi\beta) \,\&\, (\forall\beta)(\Psi\beta \supset \beta\neq A)$	8, D(=)
10.	— — — $(\exists n\beta)(\Psi\beta) \,\&\, (\forall\beta)(\Psi\beta \supset ((\Psi\beta \vee \beta=A) \,\&\, \beta\neq A))$	9
11.	— — — $(\exists n\beta)((\Psi\beta \vee \beta=A) \,\&\, \beta\neq A)$	10, Lemma 1
12.	— — — $(\Psi A \vee A=A) \,\&\, (\exists n\beta)((\Psi\beta \vee \beta=A) \,\&\, \beta\neq A)$	11, D(=)
13.	— — — $(\exists n'\beta)(\Psi\beta \vee \beta=A)$	12, D(Q′)
14.	— — — $(\forall\beta)(\Psi\beta \supset \beta\mathfrak{R}A)$	5
15.	— — — $(\forall\alpha\beta)(\alpha=A \,\&\, \Psi\beta \supset \beta\mathfrak{R}\alpha)$	14, D(=)
16.	— — — $(\forall\alpha\beta)(\alpha=A \,\&\, \Psi\beta \supset \alpha\mathfrak{R}\beta)$	15, 2
17.	— — — $(\forall\alpha\beta)(\Psi\alpha \,\&\, \beta=A \supset \alpha\mathfrak{R}\beta)$	14
18.	— — — $(\forall\alpha\beta)(\Psi\alpha \,\&\, \Psi\beta \,\&\, \alpha\neq\beta \supset \alpha\mathfrak{R}\beta)$	5
19.	— — — $(\forall\alpha\beta)(\alpha=A \,\&\, \beta=A \,\&\, \alpha\neq\beta \supset \alpha\mathfrak{R}\beta)$	D(=)
20.	— — — $(\forall\alpha\beta((\Psi\alpha \vee \alpha=A) \,\&\, (\Psi\beta \vee \beta=A) \,\&\, \alpha\neq\beta \supset \alpha\mathfrak{R}\beta)$	16–19
21.	— — — $\Phi A \,\&\, (\forall\beta)(\Psi\beta \supset \Phi\beta)$	5
22.	— — — $(\forall\beta)(\beta=A \supset \Phi\beta) \,\&\, (\forall\beta)(\Psi\beta \supset \Phi\beta)$	21, D(=)
23.	— — — $(\forall\beta)((\Psi\beta \vee \beta=A) \supset \Phi\beta)$	22
24.	— — — $(\exists\psi)((\exists n'\beta)(\psi\beta) \,\&\, (\forall\alpha\beta)(\psi\alpha \,\&\, \psi\beta \,\&\, \alpha\neq\beta \supset \alpha\mathfrak{R}\beta) \,\&\, (\forall\beta)(\psi\beta \supset \Phi\beta)$	13, 20, 23
25.	— — — $(\exists n'_{\mathfrak{R}}\,\alpha)(\Phi\alpha)$	24, D($Q_{\mathfrak{R}}$)
26.	— — $(\exists n'_{\mathfrak{R}}\,\alpha)(\Phi\alpha)$	4, 5–25
27.	— $(\exists\alpha)(\Phi\alpha \,\&\, (\exists n_{\mathfrak{R}}\,\beta)(\Phi\beta \,\&\, \beta\mathfrak{R}\alpha)) \supset (\exists n'_{\mathfrak{R}}\,\alpha)(\Phi\alpha)$	3–26
28.	$\mathrm{Count}\,(\mathfrak{R}) \supset ((\exists\alpha)(\Phi\alpha \,\&\, (\exists n_{\mathfrak{R}}\,\beta)(\Phi\beta \,\&\, \beta\mathfrak{R}\alpha)) \supset (\exists n'_{\mathfrak{R}}\,\alpha)(\Phi\alpha))$	1–27

So far our numerical letters and variables have occurred only as parts of quantifier expressions, but by making use of our last definitions we now introduce formulae containing numerical letters and variables 'in isolation', occupying contexts which are not available for quantifier letters and variables[29]

$D(=_N)$:	'$n=_N m$'	for	'$(\forall\mathfrak{R})(\exists n_{\mathfrak{R}} \equiv \exists m_{\mathfrak{R}})$'
$D(\mathbf{S}_N)$:	'$n\mathbf{S}_N m$'	for	'$(\forall\mathfrak{R})(\exists n_{\mathfrak{R}} \equiv \exists m'_{\mathfrak{R}})$'
$D({*}\mathbf{S}_N)$:	'$n{*}\mathbf{S}_N m$'	for	'$(\forall\mathfrak{R})(((\forall \mathfrak{n}\mathfrak{m})(\mathfrak{n}\mathbf{S}_N \mathfrak{m} \supset \exists\mathfrak{n}\mathfrak{R}\exists\mathfrak{m})$ & $(\forall\mathfrak{n}\mathfrak{m}\mathfrak{p})(\mathfrak{n}\,\mathbf{S}_N\mathfrak{p}$ & $\exists\mathfrak{p}\mathfrak{R}\exists\mathfrak{m} \supset \exists\mathfrak{n}\mathfrak{R}\exists\mathfrak{m})) \supset \exists\mathfrak{n}\mathfrak{R}\exists\mathfrak{m})$'

The new formulae thus introduced are, however, equivalent to the old formulae employing numerical quantifiers based on non-identity, as I now deduce. The first two deductions run quite parallel to one another.

THEOREM 0·6. $n=_N m \equiv (\exists n \equiv \exists m)$

1.	$n=_N m \supset (\forall\mathfrak{R})(\exists n_{\mathfrak{R}} \equiv \exists m_{\mathfrak{R}})$	D $(=_N)$
2.	$n=_N m \supset (\exists n_{\neq} \equiv \exists m_{\neq})$	1
3.	$n=_N m \supset (\exists n \equiv \exists m)$	2, Thm 0·1
4.	— $\exists n \equiv \exists m$	Hyp
5.	— $(\forall\psi)((\exists n\alpha)(\psi\alpha) \equiv (\exists m\alpha)(\psi\alpha))$	4, D(≡)
6.	— — $(\exists n_{\mathfrak{R}}\alpha)(\Phi\alpha)$	Hyp
7.	— — $(\exists\psi)((\exists n\alpha)(\psi\alpha)$ & $(\forall\alpha\beta)(\psi\alpha$ & $\psi\beta$ & $\alpha\neq\beta \supset \alpha\mathfrak{R}\beta)$ & $(\forall\alpha)(\psi\alpha \supset \Phi\alpha))$	6, $D(Q_{\mathfrak{R}})$
8.	— — $(\exists\psi)((\exists m\alpha)(\psi\alpha)$ & $(\forall\alpha\beta)(\psi\alpha$ & $\psi\beta$ & $\alpha\neq\beta \supset \alpha\mathfrak{R}\beta)$ & $(\forall\alpha)(\psi\alpha \supset \Phi\alpha))$	5, 7
9.	— — $(\exists m_{\mathfrak{R}}\alpha)(\Phi\alpha)$	8, $D(Q_{\mathfrak{R}})$
10.	— $(\exists n_{\mathfrak{R}}\alpha)(\Phi\alpha) \supset (\exists m_{\mathfrak{R}}\alpha)(\Phi\alpha)$	6–9
11.	— $(\exists m_{\mathfrak{R}}\alpha)(\Phi\alpha) \supset (\exists n_{\mathfrak{R}}\alpha)(\Phi\alpha)$	Similarly
12.	— $(\exists n_{\mathfrak{R}}\alpha)(\Phi\alpha) \equiv (\exists m_{\mathfrak{R}}\alpha)(\Phi\alpha)$	10, 11
13.	— $(\forall\mathfrak{R})(\forall\phi)((\exists n_{\mathfrak{R}}\alpha)(\phi\alpha) \equiv (\exists m_{\mathfrak{R}}\alpha)(\phi\alpha))$	4–12
14.	— $(\forall\mathfrak{R})(\exists n_{\mathfrak{R}} \equiv \exists m_{\mathfrak{R}})$	13, D(≡)
15.	— $n=_N m$	14, $D(=_N)$
16.	$(\exists n \equiv \exists m) \supset n=_N m$	4–15
17.	$n=_N m \equiv (\exists n \equiv \exists m)$	3, 16

THEOREM 0·7. $n\mathbf{S}_N m \equiv (\exists n\,\mathbf{S}\,\exists m)$

1.	$n\mathbf{S}_N m \supset (\forall\mathfrak{R})(\exists n_{\mathfrak{R}} \equiv \exists m'_{\mathfrak{R}})$	$D(\mathbf{S}_N)$
2.	$n\mathbf{S}_N m \supset (\exists n_{\neq} \equiv \exists m'_{\neq})$	1
3.	$n\mathbf{S}_N m \supset (\exists n \equiv \exists m')$	2, Thms 0·1 0·2
4.	$n\mathbf{S}_N m \supset (\exists n\,\mathbf{S}\,\exists m)$	3, D(**S**), D(≡)
5.	— $\exists n\,\mathbf{S}\,\exists m$	Hyp
6.	— $(\forall\psi)((\exists n\alpha)(\psi\alpha) \equiv (\exists m'\alpha)(\psi\alpha))$	5, D(**S**)
7.	— — $(\exists n_{\mathfrak{R}}\alpha)(\Phi\alpha)$	Hyp
8.	— — $(\exists\psi)((\exists n\alpha)(\psi\alpha)$ & $(\forall\alpha\beta)(\psi\alpha$ & $\psi\beta$ & $\alpha\neq\beta \supset \alpha\mathfrak{R}\beta)$ & $(\forall\alpha)(\psi\alpha \supset \Phi\alpha))$	7, $D(Q_{\mathfrak{R}})$

[29] Note that we have given no sense to the formulae '$Q_1=_N Q_2$' or '$n\equiv m$'.

9. $— — (\exists\psi)((\exists m'\alpha)(\psi\alpha) \,\&\, (\forall\alpha\beta)(\psi\alpha \,\&\, \psi\beta \,\&\, \alpha\neq\beta \supset \alpha\mathfrak{R}\beta)$
$\qquad \& \, (\forall\alpha)(\psi\alpha \supset \Phi\alpha))$ — 8, 6
10. $— — (\exists m'_{\mathfrak{R}}\alpha)(\Phi\alpha)$ — 9, D($Q_{\mathfrak{R}}$)
11. $— (\exists n_{\mathfrak{R}}\alpha)(\Phi\alpha) \supset (\exists m'_{\mathfrak{R}}\alpha)(\Phi\alpha)$ — 7–10
12. $— (\exists m'_{\mathfrak{R}}\alpha)(\Phi\alpha) \supset (\exists n_{\mathfrak{R}}\alpha)(\Phi\alpha)$ — Similarly
13. $— (\exists n_{\mathfrak{R}}\alpha)(\Phi\alpha) \equiv (\exists m'_{\mathfrak{R}}\alpha)(\Phi\alpha)$ — 11, 12
14. $— (\forall\mathfrak{R})(\forall\phi)((\exists n_{\mathfrak{R}}\alpha)(\phi\alpha) \equiv (\exists m_{\mathfrak{R}}\alpha)(\phi\alpha))$ — 5–13
15. $— (\forall\mathfrak{R})(\exists n_{\mathfrak{R}} \equiv \exists m'_{\mathfrak{R}})$ — 14, D($\equiv$)
16. $— n\mathbf{S}_N m$ — 5, D($\mathbf{S}_N$)
17. $(\exists n\, \mathbf{S}\, \exists m) \supset n\mathbf{S}_N m$ — 5–16
18. $n\mathbf{S}_N m \equiv (\exists n\, \mathbf{S}\, \exists m)$ — 4, 17

The last of these three deductions has already been prepared for in the previous appendix, for it is simply an application of the theorem there proved as A11.

THEOREM 0·8. $n{*}\mathbf{S}_N m \equiv \exists n {*}\mathbf{S} \exists m$

1. $(\forall q_1q_2)(q_1\mathbf{S}q_2 \,\&\, q_2{*}\mathbf{S}\exists \supset q_1{*}\mathbf{S}\exists)$ — A8
2. $Q_2{*}\mathbf{S}\exists \supset (Q_1{*}\mathbf{S}Q_2 \equiv (\forall\mathfrak{R})(((\forall q_1q_2)(q_1{*}\mathbf{S}\exists \,\&\, q_1\mathbf{S}q_2 \,\&\, q_2{*}\mathbf{S}\exists$
$\supset q_1\mathfrak{R}q_2) \,\&\, (\forall q_1q_2q_3)(q_1{*}\mathbf{S}\exists \,\&\, q_1\mathbf{S}q_2 \,\&\, q_2{*}\mathbf{S}\exists \,\&\, q_2\mathfrak{R}q_3 \,\&$
$q_3{*}\mathbf{S}\exists \supset q_1\mathfrak{R}q_3)) \supset Q_1\mathfrak{R}Q_2))$ — 1, A11
3. $\exists n {*}\mathbf{S} \exists m \equiv (\forall\mathfrak{R})(((\forall q_1q_2)(q_1{*}\mathbf{S}\exists \,\&\, q_1\mathbf{S}q_2 \,\&\, q_2{*}\mathbf{S}\exists \supset q_1\mathfrak{R}q_2)$
$\& \, (\forall q_1q_2q_3)(q_1{*}\mathbf{S}\exists \,\&\, q_1\mathbf{S}q_2 \,\&\, q_2{*}\mathbf{S}\exists \,\&\, q_2\mathfrak{R}q_3 \,\&\, q_3{*}\mathbf{S}\exists \supset$
$q_1\mathfrak{R}q_3)) \supset \exists n\mathfrak{R}\exists m)$ — 2, D(n)
4. $\exists n {*}\mathbf{S} \exists n \equiv (\forall\mathfrak{R})((\forall \mathfrak{n}\mathfrak{m})(\exists\mathfrak{n}\mathbf{S}\exists\mathfrak{m} \supset \exists\mathfrak{n}\mathfrak{R}\exists\mathfrak{m}) \,\&$
$(\forall\mathfrak{n}\mathfrak{m}\mathfrak{p})(\exists\mathfrak{n}\mathbf{S}\exists\mathfrak{m} \,\&\, \exists\mathfrak{m}\mathfrak{R}\exists\mathfrak{p} \supset \exists\mathfrak{n}\mathfrak{R}\exists\mathfrak{p})) \supset \exists n\mathfrak{R}\exists m)$ — 3, D($\mathfrak{n}$)
5. $\exists n {*}\mathbf{S} \exists m \equiv (\forall\mathfrak{R})(((\forall\mathfrak{n}\mathfrak{m})(\mathfrak{n}\mathbf{S}_N\,\mathfrak{m} \supset \exists\mathfrak{n}\mathfrak{R}\exists\mathfrak{m}) \,\&$
$(\forall\mathfrak{n}\mathfrak{m}\mathfrak{p})(\mathfrak{n}\mathbf{S}_N\,\mathfrak{m} \,\&\, \exists\mathfrak{m}\mathfrak{R}\exists\mathfrak{p} \supset \exists\mathfrak{n}\mathfrak{R}\exists\mathfrak{p})) \supset \exists n\mathfrak{R}\exists m)$ — 4, Thm 0·7
6. $\exists n {*}\mathbf{S} \exists m \equiv n{*}\mathbf{S}_N m$ — 5, D($*\mathbf{S}_N$)

Let us now call the formulae '$n=_N m$', '$n\,\mathbf{S}_N m$', and '$n{*}\mathbf{S}_N m$' (and their alphabetic variants) the elementary *arithmetical* formulae, and '$\exists n \equiv \exists m$', '$\exists n\, \mathbf{S}\, \exists m$', and '$\exists n {*}\mathbf{S}\, \exists m$' (and their alphabetic variants) the *corresponding quantifier* formulae. Theorems 0·6–0·8 then constitute a proof that any elementary arithmetical formula is equivalent to its corresponding quantifier formula, and from this it follows that any more complex arithmetical formula built up from elementary arithmetical formulae by truth-functional combination and universal quantification will be equivalent to its more complex corresponding quantifier formula. For, if '(— ∃n —)', '(... ∃n ...)', and '(-- ∃m --)' are any formulae containing the quantifiers '∃n' and '∃m', then it is an entirely trivial matter to establish, from D(n) and D($\mathfrak{n}$), that

(i) If $\vdash (— \exists n —) \equiv (\ldots \exists n \ldots)$
then $\vdash {\sim}(— \exists n —) \equiv {\sim}(\ldots \exists n \ldots)$

(ii) If $\vdash (— \exists n —) \equiv (\ldots \exists n \ldots)$
then $\vdash ((— \exists n —) \,\&\, (-- \exists m --)) \equiv ((\ldots \exists n \ldots) \,\&\, (-- \exists m --))$

(iii) If $\vdash (\text{—}\exists\text{n}\text{—}) \equiv (\ldots\exists\text{n}\ldots)$
then $\vdash (\forall\mathfrak{n})(\text{—}\exists\mathfrak{n}\text{—}) \equiv (\forall\mathfrak{n})(\ldots\exists\mathfrak{n}\ldots)$

and hence our general metatheorem follows by induction. For we may take '$(\text{—}\exists\text{n}\text{—})$' as the quantifier formula corresponding to any desired arithmetical formula and '$(\ldots\exists\text{n}\ldots)$' as the definitional expansion of that arithmetical formula in terms of the initial numerical quantifiers. And, finally, we may extend this result to cover also formulae built up by applying the number-quantifier '$\exists\text{n}_{\neq_N}$' in view of the fact that

(iv) If $\vdash (\text{—}\exists\text{n}\text{—}) \equiv (\ldots\exists\text{n}\ldots)$
then $\vdash (\exists\text{m}_{\neq_N}\mathfrak{n})(\text{—}\exists\mathfrak{n}\text{—}) \equiv (\exists\text{m}_{\neq_N}\mathfrak{n})(\ldots\exists\mathfrak{n}\ldots)$.

It only remains to give a proof of this last assertion, and we will then have shown that arithmetical formulae may be exchanged for their corresponding quantifier formulae in any of the contexts we need to consider when deriving the fundamental theses of arithmetic.

It will be recalled that the quantifier '$\exists\text{n}_{\neq_N}$' is defined for numerical variables (and for them alone) by the definition

$$D(\exists\text{n}_{\neq_N})\text{: }`(\exists\text{n}_{\neq_N}\mathfrak{m})(\Phi(\exists\mathfrak{m}))\text{'} \quad \text{for} \quad `(\exists\text{n}_{\neq}\text{q})(\Phi\text{q} \ \&\ \text{q}{*}\mathbf{S}\exists)\text{'}.$$

With this definition the proof required may be given very simply thus:

THEOREM 0·9. $(\forall\mathfrak{m})(\Phi(\exists\mathfrak{m})\supset\Psi(\exists\mathfrak{m}))\supset ((\exists\text{n}_{\neq_N}\mathfrak{m})(\Phi(\exists\mathfrak{m}))\supset(\exists\text{n}_{\neq_N}\mathfrak{m})(\Psi(\exists\mathfrak{m}))$

1.	— $(\forall\alpha)(\Phi\alpha\supset\Psi\alpha)$	Hyp
2.	— — $(\exists\text{n}_{\mathfrak{R}}\alpha)(\Phi\alpha)$	Hyp
3.	— — $(\exists\psi)((\exists\text{n}\alpha)(\psi\alpha) \ \&\ (\forall\alpha\beta)(\psi\alpha \ \&\ \psi\beta \ \&\ \alpha\neq\beta\supset\alpha\mathfrak{R}\beta) \ \&\ (\forall\alpha)(\psi\alpha\supset\Phi\alpha))$	2, D(Q$_{\mathfrak{R}}$)
4.	— — $(\exists\psi)((\exists\text{n}\alpha)(\psi\alpha) \ \&\ (\forall\alpha\beta)(\psi\alpha \ \&\ \psi\beta \ \&\ \alpha\neq\beta\supset\alpha\mathfrak{R}\beta) \ \&\ (\forall\alpha)(\psi\alpha\supset\Psi\alpha))$	1, 3
5.	— — $(\exists\text{n}_{\mathfrak{R}}\alpha)(\Psi\alpha)$	4, D(Q$_{\mathfrak{R}}$)
6.	— $(\exists\text{n}_{\mathfrak{R}}\alpha)(\Phi\alpha)\supset(\exists\text{n}_{\mathfrak{R}}\alpha)(\Psi\alpha)$	2–5
7.	$(\forall\alpha)(\Phi\alpha\supset\Psi\alpha)\supset((\exists\text{n}_{\mathfrak{R}}\alpha)(\Phi\alpha)\supset(\exists\text{n}_{\mathfrak{R}}\alpha)(\Psi\alpha))$	1–6
8.	$(\forall\text{q})(\Phi\text{q}\supset\Psi\text{q})\supset((\exists\text{n}_{\mathfrak{R}}\text{q})(\Phi\text{q})\supset(\exists\text{n}_{\mathfrak{R}}\text{q})(\Psi\text{q}))$	7
9.	$(\forall\text{q})(\Phi\text{q}\supset\Psi\text{q})\supset((\exists\text{n}_{\neq}\text{q})(\Phi\text{q})\supset(\exists\text{n}_{\neq}\text{q})(\Psi\text{q}))$	8
10.	$(\forall\text{q})((\Phi\text{q} \ \&\ \text{q}{*}\mathbf{S}\exists)\supset(\Psi\text{q} \ \&\ \text{q}{*}\mathbf{S}\exists)) \supset((\exists\text{n}_{\neq}\text{q})(\Phi\text{q} \ \&\ \text{q}{*}\mathbf{S}\exists)\supset(\exists\text{n}_{\neq}\text{q})(\Psi\text{q} \ \&\ \text{q}{*}\mathbf{S}\exists))$	9
11.	$(\forall\text{q})(\text{q}{*}\mathbf{S}\exists\supset(\Phi\text{q}\supset\Psi\text{q})) \supset((\exists\text{n}_{\neq}\text{q})(\Phi\text{q} \ \&\ \text{q}{*}\mathbf{S}\exists)\supset(\exists\text{n}_{\neq}\text{q})(\Psi\text{q} \ \&\ \text{q}{*}\mathbf{S}\exists))$	10
12.	$(\forall\mathfrak{m})(\Phi(\exists\mathfrak{m})\supset\Psi(\exists\mathfrak{m})) \supset((\exists\text{n}_{\neq}\text{q})(\Phi\text{q} \ \&\ \text{q}{*}\mathbf{S}\exists)\supset(\exists\text{n}_{\neq}\text{q})(\Psi\text{q} \ \&\ \text{q}{*}\mathbf{S}\exists))$	11, D($\mathfrak{n}$)
13.	$(\forall\mathfrak{m})(\Phi(\exists\mathfrak{m})\supset\Psi(\exists\mathfrak{m})) \supset((\exists\text{n}_{\neq_N}\mathfrak{m})(\Phi(\exists\mathfrak{m}))\supset(\exists\text{n}_{\neq_N}\mathfrak{m})(\Psi(\exists\mathfrak{m})))$	12, D($\exists\text{n}_{\neq_N}$)

6: Logicism

1. Logic

IN the preceding chapters I have produced a construction of arithmetic which appears to be a complete vindication of the logicist claim, for the logicist claim is that the truths of arithmetic may be deduced from truths of pure logic with the sole aid of definitions, and this—it would seem—is something that I have now done. There are, however, some important objections to be met before we can rest content with this conclusion, and it is now time to consider them. The objections fall naturally under two headings: is it in fact true that the basis from which I start is a basis of pure logic, and is it in fact true that what results from my construction is arithmetic?

The basis for my construction is of course not the same as the basis that Frege originally proposed for his construction when he first propounded the logicist claim; and the major difference between us is that whereas Frege employed an analysis of propositions into subjects and predicates, making only small use of his second-level predicates (which I call subject-quantifiers), I rely rather on an analysis of propositions into quantifiers and propositional forms which results in a rather different kind of system. Both of us of course employ the ordinary logic of truth-functions embodied in the propositional calculus, and the logic of first-order quantification theory in application to our respective propositional components, but there are considerable differences in the substitution-rules we adopt in higher-order quantification theory.[1] Frege was working within the lower levels of the orthodox hierarchy of types, and so could use the two simple 'remainder' principles that any proposition of any form '(—a—)' is also of the related form 'Fa' and that any proposition

[1] By 'first-order' quantification theory I mean the theory given in chapter 3, section 2, by rules (i)–(iv) with (iv)*a*. By 'higher-order' quantification theories I mean the theories which result by adding further specifications of rule (iv), which will then function as further rules of substitution.

of any form '(—F—)' is also of the related form '(𝔔x)(Fx)'; but that was all. My system is rather more complex, but its most significant features could perhaps be summed up like this. Analogous to Frege's two remainder principles are the principles that any proposition of any form '(—A—)' is also of the related form 'ΦA' and that any proposition of any form '(—Φ—)' is also of the related form '(𝔔α)(Φα)', but although I do use the first of these the second figures only in a restricted form: in fact the principle I use is that propositions of *some* of the forms '(—Φ—)', namely those forms in which 'Φ' is contained 'type-neutrally', are also of the related pure-quantifier form '(Qα)(Φα)'. But though my system is more limited in this way it is in another way much more powerful, because of the extra substitution rule which I call substitution for α-symbols. My use of this latter rule could be roughly summed up as absorbing the principle that any proposition of any form '(—Q—)' is also of the related form 'ΦQ', and then adding the principle that any proposition of any form '(—ΦQ—)' is also of the (non-related) form '(—ΦA—)'. It is essentially these further principles, together with the conception of a pure quantifier as capable of occurring both in the context '(Qα)(Φα)' and in the context '(Qq)(Φq)', that enable me to achieve my end without at any point invoking Frege's disastrous fifth axiom, which introduced classes in a way leading straight to Russell's paradox.

Frege's logic without his fifth axiom is presumably not powerful enough to serve as a basis for the deduction of arithmetic, and I think that the same is true of other systems which one could reasonably regard as 'purely logical', with the possible exception of Borkowski's. If, like Russell, one remains within the framework of the orthodox hierarchy of types, then there seems no way of avoiding the postulation of an axiom of infinity as an extra non-logical axiom; while if, like Quine, one avoids *this* axiom by employing a theory of sets with sufficiently comprehensive principles of set existence, then these principles of set existence themselves seem very doubtful as truths of logic. At any rate no one, so far as I know, has even attempted to justify sufficiently comprehensive principles of set existence as a consequence of an analysis of the notion of a set, as I have attempted to justify the principles of the logic of quantifiers by an analysis of the notion of a quantifier.

As for Borkowski's system, that differs from mine mainly in that it makes no use of α-symbols but is entirely constructed from (his analogue to my) Q-symbols and Φ-symbols.[2] If my α-symbols are replaced everywhere by Q-symbols—so that my Φ-symbols come to represent only predicates of quantifiers—this would not affect the validity of my deductions in my own system (though of course it would affect their meaning), and the deductions would then be valid also in Borkowski's system. But the converse does not hold, for Borkowski's deductions would occasionally fail in my system. Although he does not give his substitution-rules very explicitly, it would appear that Borkowski shares with me the principle that any proposition of any form '(—Q—)' is also of the related form 'ΦQ', thus legitimising substitution of Φ-schemata for Φ-letters; but he goes beyond me in assuming quite generally that every proposition of any form '(—Φ—)' is also of the related form '(Qq)(Φq)', thus allowing himself to substitute any general $\mathfrak{Q}$q-schema for an elementary schema '(Qq)(—q—)'. At any rate he not only assumes in his D6 that '∼(∃q)(—q—)' is a legitimate substitution for '(Qq)(—q—)', but in his D3 he also assumes the same of the somewhat complicated schema

$$(\exists\mathfrak{R})(\text{one-to-one}(\mathfrak{R})\ \&\ (\forall q_1)(\Phi q_1 \supset (\exists q_2)(— q_2 —\ \&\ q_1\mathfrak{R}q_2))\ \&\ (\forall q_2)(— q_2 — \supset (\exists q_1)(\Phi q_1\ \&\ q_1\mathfrak{R}q_2))).$$

However, this latter schema would *not* be permitted in my system as a substitution for the elementary schema '(Qq)(—q—)', since it is *not* a pure Q-schema.[3] In fact I am not convinced that Borkowski's system does have an interpretation in which all his deductions are acceptable, but if it does it seems most likely that it will turn out to be some fragment of the general logic of quantifiers that I sketched in the appendix to chapter 4, and in that case the objections that I should wish to raise against it would be of the more philosophical sort canvassed on pp. 18–20.

[2] I am disregarding his anomalous use of 'typically ambiguous' symbols in his D6, on the grounds given in note 14 to chapter 1. In view of my somewhat extended discussion in chapter 4 (section 4 and Appendix), it will be evident that I do not find his brief note on this topic entirely satisfactory (note 7, p. 259). I also disregard his superfluous axiom of extensionality.

[3] Yet Borkowski's Q-symbols are *intended* by him to represent only the pure quantifiers, for he says 'we assume the principle of the independence of the semantic category of the quantifier on the semantic category of the quantifier-variable' (p. 283).

Anyway, if we may suppose that Borkowski's system can (if sound) be assimilated to mine, it would seem fair to say that the logical system I put forward is more powerful than other known systems which could claim to be 'purely logical', just because it *is* adequate as a basis for the construction of elementary arithmetic, but I do not see that it must for that reason be denied the title of 'pure logic' itself. Since we lack any agreed criterion for what does count as logic this claim is not one that is capable of being established or refuted in a definite way (provided of course that my system is consistent), but it is not very clear to me what grounds one might plausibly have for objecting to it. There is no very evident reason for holding that the fundamental analysis of propositions into quantifiers and propositional forms is any less a matter of 'logic' than is the usual analysis into subjects and predicates, while the principles of truth-functional logic and first-order quantification theory are presumably no more 'logical' in one application than in another, so the debate might reasonably be expected to centre round the extra substitution rules that distinguish higher-order theories. But in fact a good deal of the debate has concerned itself rather with the application of *first*-order quantification theory to variables other than subject-variables, with suspicion being directed at any formula involving quantification even over predicates, so perhaps I should begin by recalling my argument on this point.

In chapter 3 I discussed Quine's view that quantification which really is quantification *over predicates* makes no sense, because the formulae involved have no appropriate English rendering. The consequence Quine drew from this view was that quantification over predicates must really be interpreted as quantification over sets, and if that were indeed so then I should not wish to claim it as a part of 'pure logic'. (For, as I have already indicated, I find it difficult to regard as pure logic any of the several competing theories concerning the existence of sets.) But this objection has I hope been dispelled by my earlier discussion, where I proposed a way of reading these quantified formulae which does not seem open to Quine's objections and which does justify the usual rules and axioms for them. The same objection, that quantification over predicates makes no sense, is sometimes advanced on the somewhat different ground that it presupposes the existence of 'a precise and well-defined totality of all predicates', whereas in fact there is no such totality.[4] An initial rejoinder to this objection might be that if it is correct then presumably quantification over objects must likewise presuppose the existence of 'a precise and well-defined totality of all objects', but what reason is there to suppose that this second totality does exist while the first does not?[5] But I do not think it helps to speak here of 'precise and well-defined totalities', for presumably the point being made is just that we must know what counts as a predicate before we can meaningfully speak of all predicates, and similarly that we must know what counts as an object before we can meaningfully speak of all objects. But my reply to this is first that the knowledge desired is in fact needed even earlier if we are to understand the use of the relevant schematic letters to indicate generality, though the critic does not usually raise any objection to this, and second that it is anyway not so much the notion of an object or a predicate that we must understand as the notion of a proposition which contains a subject, or a proposition which contains a predicate, and similar notions. At this point the critic might perhaps admit that if the notion of a proposition was clear then there might be no insuperable difficulty in the notion of a proposition which contains a predicate—or at any rate no more than there is in the notion of a proposition which contains a subject—but, he may go on, the notion of a proposition simply is not clear, and that is where the difficulty lies. Put in this form, the objection is one which I make no attempt to answer in this book. I have already argued that we cannot understand even ordinary first-order predicate logic in its relation to propositional (or so-called 'sentential') logic without presupposing this notion, and so the objection is now no longer an objection to any facet of higher-order theories but one that clearly affects even the most elementary stages of logic. And ultimately I think that the whole edifice of classical two-valued logic depends upon the classical

[4] H. Putnam, 'The Thesis that Mathematics is Logic', section 1 (in *Bertrand Russell, Philosopher of the Century*, ed. R. Schoenman). Compare e.g. C. Parsons, 'Frege's Theory of Number' (in *Philosophy in America*, ed. M. Black).

[5] One is perhaps tempted to think the totality of all objects less mysterious than the totality of all predicates because predicates appear to be dependent upon (possible) languages in a way that objects do not. I think myself that this appearance is delusive, but the point is one that I can only argue *ad hominem*, when faced with an opponent who is prepared to explain his conception of an object, and consequently I do not attempt it here.

concept of a proposition, and it is this which is the most fundamental assumption in the logicist approach. But there I will have to leave the matter.

Returning, then, to other objections that are brought against higher-order theories, in several cases it seems to be the *complexity* of these theories that is the stumbling-block. But I cannot see that that has anything to do with the matter. In a fairly obvious sense, higher-order theories are indeed more complicated than ordinary first-order predicate calculus, and that in turn is more complicated than propositional calculus, and that again is more complicated than Aristotle's syllogistic; but surely no one would wish to claim that this scale of decreasing complexity is also a scale of increasing logical purity? On the contrary it seems more plausible to argue that the logician's desire for generality is what leads him to *advance* from the very simple but very limited results of Aristotelian syllogistic to the more general, and so more abstract, systems that are now orthodox; and clearly systems that are more general and more abstract should be expected to be more complicated. But in fact I suspect that it is not so much the bare complexity of higher-order theories that accounts for their being denied the title of 'logic', but rather our ignorance about them. Whereas the metatheory of first-order predicate calculus is now very well known, with higher-order theories we very often have no proof even of their consistency, and we know that we shall need all the mathematician's skills to settle such matters for us.[6] But obviously we cannot validly draw any conclusions simply from ignorance, and besides it seems to me that the techniques one needs to establish the metatheory of a given system are quite irrelevant to the question whether that system counts as a purely logical one; *that* is a matter of the concepts that are taken as primitive in the system, and the nature of the rules and axioms laid down for them. There is, however, one specific metatheoretical feature of higher-order theories that has been held against them, and in particular has been used to put even second-order predicate logic beyond the pale of logic properly so-called, and that is their lack of *completeness*.[7] It will be worth our while to consider this point somewhat more fully.

[6] Just as we once needed the mathematician's skills to determine, e.g., that first-order predicate calculus lacks a decision procedure.

[7] See for instance W. & M. Kneale, *The Development of Logic*, chapter XIII, section 5.

The completeness of first-order predicate calculus consists in the fact that it cannot properly be extended by adding as a new axiom some formula *which is already a formula of the calculus* but is not already provable. Any such addition must take us beyond the realm of logical truth, since we would then be adding as an axiom a formula which comes out false under some interpretation in some possible universe. However, this by itself cannot be taken to imply that the truths of first-order predicate calculus are all the logical truths there are, because of course we could say exactly the same thing about *propositional* calculus. It is equally true that propositional calculus cannot properly be extended by adding as a new axiom some formula which is already a formula of the calculus but is not already provable—indeed in this case the addition must lead to actual inconsistency—but I take it that no one will be willing to conclude from this that propositional calculus already contains *all* logical truths. Obviously we can add new logical truths to the basis provided by propositional calculus by adding *new formulae* with appropriate rules and axioms to govern them, for instance the formulae of first-order predicate calculus, and I do not see why we should not then apply exactly the same procedure once again. Indeed I think it would be quite commonly admitted that we do not exceed the bounds of logic if we add to first-order predicate calculus formulae involving the new sign '=' for Leibnizian identity, with appropriate axioms, and a case might well be made for proceeding yet further in this way (for instance by adding new formulae and axioms for the ancestral). But the addition of the new formulae of second-order predicate calculus is even less of a new departure than the addition of an identity sign, for (as I argued in chapter 3, section 1) no new *idea* is thereby being introduced. As for the new rules or axioms to be adopted with the new formulae, in the case of second-order predicate calculus I would only claim the rule (iv)*b* as a truth of logic, and any doubts one might have about this ought to be stilled by the reflection that it is in fact already presupposed in our operation of the first-order theory. Indeed if the first-order theory is presented with a finite number of axioms, and consequently a rule of substitution, then this rule of substitution will precisely be the rule (iv)*b* which is therefore explicitly present from the start. On the other hand if the first-order theory is presented with axiom-*schemata*, each generating infinitely many axioms, then we are in effect already employing rule (iv)*b* when we do generate the axioms from the axiom-schemata, and so again we have in a way already acknowledged this rule. There is a sense, therefore, in which we are deliberately refusing to acknowledge all our premisses if we maintain that the first-order theory is logic while the second-order theory is not.

Now it is one of the consequences of Gödel's work on the incompleteability of formal arithmetic that even this second-order predicate calculus is essentially incomplete—and the same must certainly be true of my 'logic of pure quantifiers'—but it seems to me that the correct moral to draw from this is not that higher-order quantification theory fails to be logic but rather that no strictly formalized system of logic can contain all the logical truths there are. Nor is this conclusion at all a surprising one. For in order to be a 'strictly formalized system' in the sense required for the application of Gödel's theorem it is necessary that the system be presented with a fixed and finite vocabulary of primitive symbols, and recursively specifiable rules for determining what counts as a well-formed formula of the system and what counts as a proof. The whole strategy of Gödel's argument is then to operate in an enlarged system, with a further vocabulary (of 'protosyntax'), so as to be able to express more than could be expressed in the original system, and in this way he is able to show that a certain unprovable formula of the original system is nevertheless true (in its standard interpretation). It seems reasonable to maintain, therefore, that there is a perfectly comprehensible reason why no formal system can contain all the logical truths, namely that the reasoning that can be carried out *within* a strictly formalized system is always limited by its vocabulary and syntax. We, however, are always capable of going outside that system and reasoning with an enlarged vocabulary, and—in suitable cases—our reasoning with the enlarged vocabulary will issue in conclusions that can be stated with the original and smaller vocabulary but not proved within the original system. However, I would be very loath to take this limitation on what can be done with strictly formalized systems as a limitation on the sort of reasoning that can properly be called 'logical', and hence as a limitation on what counts as a logical truth.

This rather summary discussion of Gödel's theorem brings me on to a version of the second objection which I said deserved consideration, namely the objection that whatever it is that we have deduced it is not (the whole of) *arithmetic*. I have been concerned only to deduce Peano's postulates for arithmetic, on the ground that once these are to hand there are well-known methods of defining all further concepts of elementary arithmetic and deducing the standard theorems for them,[8] but now it might be objected that Peano's postulates cannot provide an adequate foundation for arithmetic since Gödel's results show that there must be arithmetical truths which cannot be derived from them in any (consistent) formal system. What I have been trying to show, however, is that the inadequacy in question is not an inadequacy in Peano's postulates at all, but rather in the underlying logical system; and it is an inadequacy which any strictly formalized logical basis must share. Peano's postulates, after all, are known to be categorical,[9] which means that any two models that satisfy them must be isomorphic with one another in respect of the relations of identity and succession, and from this it follows that all the truths concerning numbers that can be formulated in terms of these two relations—i.e. all arithmetical truths—are consequences of those postulates. Certainly when I speak of *consequences* here I must not be understood to mean that they are consequences *derivable in some particular formal system*, for it is just this that is refuted by Gödel's theorem; I mean rather that provided the postulates do hold then the 'consequences' I mention must hold also, whether or not we can actually derive them within the system we are currently employing. I would rather conclude, then, that Peano's postulates do form an adequate characterization of the number series so far as arithmetical propositions are concerned, and if we have genuinely succeeded in deducing Peano's postulates then no further information about numbers is required before we can say that we have produced a sufficient foundation for arithmetic. Arithmetic will be developed from this foundation by applying reasoning which will doubtless deserve to be

[8] The general method employed by Quine in *Mathematical Logic*, chapter 6, sections 48–9, seems to me philosophically appropriate. But some reservations will emerge from the discussion of definitions in the next section.

[9] See e.g. Church, *Introduction to Mathematical Logic*, part I, chapter V, section 55.

called 'logical', but we should not suppose that *all* this reasoning can be encapsulated in some *one* strictly formal system.

So far I have offered some defence of the claim that the logical basis I provide does deserve to be accounted a basis of 'pure logic', and I have tried to explain why I do not consider the logicist thesis to be overthrown by the incompleteability of formal arithmetic. This last point certainly commits me to a somewhat open-ended concept of 'logic', and some may wish to reject the logicist thesis on this score; but I think that in fact there are more interesting difficulties still outstanding. For I have said that I will count the logicist thesis as established if we have genuinely succeeded in deducing Peano's postulates, and by this I mean to imply that it must actually be *Peano's postulates* that we have deduced and not some other theses which, though they may have the same formal structure as these, have an entirely different *meaning*. And this question evidently leads us to consider more closely the role of the *definitions* that must figure in any logicist construction.

2. Definition and Analysis

In chapter 2, section 1, I remarked that we may distinguish at least two ways of regarding a definition: it may be taken simply as the introduction of a new *sign* for purposes of abbreviation, or it may be understood rather as the analysis of a *concept* with which we are already familiar. Now if we understand the logicist's claim to be the claim that the truths of arithmetic just *are* truths of logic, it would seem necessary to take the definitions he employs in the first way, simply as devices which enable him to set down in an economical fashion the 'purely logical' propositions which might cover several pages if fully written out in the standard notation. But in that case his claim really makes no sense. For if that is the only role which the definitions play in his argument he has absolutely no right to conclude that these propositions are indeed propositions *of arithmetic*; they may have been presented in a notation which makes them *look* like propositions of arithmetic, but that will be all that we can say. To make sense of the logicist's claim, then, it would seem that we must regard his definitions as providing *analyses* of the relevant arithmetical concepts (and other auxiliary concepts, such as that of the ancestral), and therefore if his claim is to succeed the definitions that he offers will presumably have to be granted to be 'correct' analyses. But it certainly is not easy to find a satisfactory criterion for what counts as a 'correct' analysis.

When a definition is put forward it is usual, if the definition is at all an important one, to provide some ground for thinking that the definition is 'adequate' to the concept being defined, and I did generally follow this procedure in the last chapter. We might therefore expect the question of the 'correctness' of an analysis to be the same as the question of the 'adequacy' of a definition, but I think that this expectation is on the whole mistaken. For an argument for the adequacy of a definition is generally just an argument for the *truth* of the resulting equivalence (or perhaps for its necessary truth), whereas for the definition to constitute a correct analysis it would seem to be required that definiens and definiendum actually have *the same meaning*, which is altogether a stronger requirement. For what must apparently be maintained is, for instance, that the sign '... *R ...' as we have defined it does indeed *mean* what we understand by the ancestral of the relation '... R ...', and this is surely not established by demonstrating that the one applies wherever and only where the other does. Let us consider this example in a little more detail.

It seems to me that what we actually mean by the (proper) ancestral of the relation '... R ...' can perfectly well be explained, as I did explain it, by giving the infinite disjunction

$$\text{`}R \vee R|R \vee R|R|R \vee \ldots\text{'}$$

—where the concluding dots correspond to an 'and so on'. When I came to construct a definition, however, I first offered informal arguments to show that if this is the meaning of the sign '*R' it will at least be true that

$$\vdash (R \vee R|{*}R) \subset {*}R$$
$$\vdash ((R \vee R|S) \subset S) \supset ({*}R \subset S)$$

and from this I was able to prove the truth of the equivalence

$$\vdash a{*}Rb \equiv (\forall s)(((R \vee R|s) \subset s) \supset asb)$$

which in turn I took to establish the adequacy of the definition

$$\text{`}a{*}Rb\text{'} \quad \text{for} \quad \text{`}(\forall s)(((R \vee R|s) \subset s) \supset asb)\text{'}.$$

But I certainly would not want to claim that this definiens has *the same meaning* as the original infinite disjunction. Sameness of meaning is a notoriously slippery concept to handle, and I own that I do not have any neat criterion to offer but can only appeal to what I imagine most people would find reasonable. I suppose someone *might* claim that he does just 'see' the universally quantified definiens as having the same meaning as the infinite disjunction we standardly express with 'and so on', and I do not have any general method of showing that such a person must be wrong. But most of us certainly do not just 'see' this alleged sameness of meaning; we need to assure ourselves by argument that the two are indeed equivalent, and of course it is only this that our argument does show.

However, in the end it turns out not to matter very much to the present objection whether we allow or reject a claim to sameness of meaning here, for in either case if the logicist is to hold that the result of his deductions really is arithmetic then he *must* maintain the truth of the thesis.[10]

$$\vdash a{*}Rb \equiv (\forall s)(((R \vee R|s) \subset s) \supset asb)$$

(where '$*R$' is taken as having the force of an infinite disjunction); and this remains an effective *premiss* to his deduction even if it is claimed—implausibly—that the premiss is a trivially obvious one. Since the thesis *is* an effective premiss to the deduction there is surely nothing to be gained from a philosophical point of view by the decision to *call* it a definition rather than an axiom, and I would therefore prefer to recognize it—or better, the two slightly simpler theses from which it is deduced—explicitly as axioms in the deduction. I do not want to say that there is no point at all in distinguishing between axioms and definitions, for the distinction is certainly helpful for *some* purposes. For instance, if we were engaged upon a metalogical investigation of the consistency of the system I put forward then this definition could very properly be ignored, for in *this* investigation we could indeed view it simply as a convenient abbreviation, since it does provide a rule for eliminating the new sign '$*$' from every context in which it occurs and it is obviously entirely conservative (i.e. it does not enable us to deduce any new theorems except those containing the new sign). But the point

[10] Note that I employed the infinite disjunction in introducing my *next* definition, the definition of '$(\exists n\alpha)(\Phi\alpha)$'. It seems clear that such an infinite disjunction must be introduced somewhere, for we simply do understand the natural numbers as being the numbers in the series 1, 2, 3, 4, 5, *and so on.*

is that it is not this sort of investigation that we are now pursuing; we are asking whether it is genuinely arithmetic that our deductions result in, and if it is then this 'definition' is a very important *premiss* to the deductions.[11]

In that case the question would seem to arise whether the definition, or rather the axiom it has now become, can claim to be a truth of logic. But at this point it seems to me that the question of what is and what is not a matter of 'logic' has become fruitless. The argument I have just been developing with the ancestral can of course be applied to any proffered definition of an already familiar concept, even to the definitions of the truth-functional connectives or of the existential quantifier in terms of the universal, and in these cases it seems to be generally accepted that the 'axioms' thereby introduced do count as logical truths. The case of identity is perhaps more disputed, and I judge that most people would not at once extend this title to a set of axioms for the ancestral, even though it is a concept of great generality and even though inductive arguments prove useful in reasoning about the most diverse subject-matters. But it is difficult to discern any general principles which we could use to guide our intuitive judgements in this area, and there seems little point, therefore, in insisting on one answer rather than the other. Since the axiom is justified by an *analysis* of '$*R$' as an infinite disjunction, together with some reasoning about infinite disjunctions which despite its informality is very clearly *a priori*, it seems perfectly reasonable to claim the axiom as an *analytical* truth, but that is perhaps all that can usefully be said on the matter.

Of the other definitions I have put forward the ones most deserving comment seem to be the contextual definition of the form '$(\exists n\alpha)(\Phi\alpha)$', the subsequent definition of the more general form '$(\exists n_{\mathfrak{R}}\alpha)(\Phi\alpha)$', and finally the definition of the arithmetical formula '$n =_N m$' (which carries with it the definition of '$n\mathbf{S}_N m$'). In all cases, of course, the same argument shows that these definitions do function effectively as premisses if we are claiming that the result of our deductions is arithmetic, and there

[11] Quine and Goodman have propounded a general method for dispensing with any (non-logical) axiom in favour of a definition (*Journal of Symbolic Logic*, 1940, pp. 104–9). The present argument may be taken to show that the possibility of this is in no way surprising, since for *most* purposes a definition has effectively the status of an axiom.

is therefore no philosophical loss in setting them down explicitly as axioms. Indeed in the first two cases this yields a welcome gain in perspicuity, for the attempt to present our analyses in the form of definitions in the first sense, i.e. as abbreviating conventions which allow us always to eliminate the new sign, has led to some rather artificial manœuvres.

When discussing the form '$(\exists n\alpha)(\Phi\alpha)$' I did not attempt to give an account of what propositions of this form mean, since I believe that different propositions of this form should be analysed in different ways: an analysis appropriate to a proposition of the form '$(\exists 4\alpha)(\Phi\alpha)$' need not be supposed to apply in the same way to a proposition of the form '$(\exists 400\alpha)(\Phi\alpha)$', or even to a proposition of the form '$(\exists\, 2\times 2\, \alpha)(\Phi\alpha)$'. I did, however, give an account of what is meant by saying that a proposition is of the form '$(\exists n\alpha)(\Phi\alpha)$', and this account made it clear that the opening symbol '$\exists n$' was functioning here as a schematic symbol for a pure quantifier. Accordingly the natural procedure would have been to recognize this by adding the new symbol at once to our primitive notation, giving it the status of a pure quantifier-letter, and invoking the explanation just given to justify a new axiom for the new formulae

$$\vdash (\forall \mathfrak{n})(\Phi(\exists \mathfrak{n})) \equiv (\forall q)(q * \mathbf{S}\exists \supset \Phi q).$$

This would be all that was required if we had already resolved that '$\exists n$' and '$\exists \mathfrak{n}$' counted as a genuine schematic letter and variable, and could be handled as such without more ado. Instead of this I did in fact proceed in a piecemeal way, defining first one special occurrence of the new symbol and then another, since this was necessary if we were to show that the symbol could always be seen as yielding a mere abbreviation and so eliminated from any context in which it actually occurred. But now that we have seen that the fact that this *can* be done is of only peripheral interest to the main logicist thesis, we may as well dispense with the rather cumbersome manœuvres it entails and employ the simpler and less artificial procedure just sketched.[12]

[12] This would have the consequence that the formula '$(Q\mathfrak{n})(\Phi(\exists \mathfrak{n}))$' would become admissible, for it is clear that any actual proposition of this form, where we are given a genuine quantifier in place of the schematic letter 'Q', would be already furnished with an explanation by our account of the form '$(\exists n\alpha)(\Phi\alpha)$'. But we can only give a *rule* for eliminating the numerical variable where we are already given a genuine quantifier and not a schematic quantifier letter. See p. 175.

As for the form '$(\exists n_{\mathfrak{N}}\alpha)(\Phi\alpha)$', the quest for an eliminating definition has led in this case to a different sort of artificiality, namely the employment of the two notations '$\exists n$' and '$\exists n_{\neq}$' for what is clearly intended as the same quantifier. The most natural procedure is clearly to stick to the same notation '$\exists n_{\neq}$' throughout, and to recognize this from the start as a special case of the schema '$\exists n_{\mathfrak{N}}$', but in that case our definition of this schema would have to carry with it a *second*—and circular—definition of '$\exists n_{\neq}$'. Here too the use of an axiom rather than a definition will represent a gain in the naturalness of our formulae and at the same time a clarification of our claim to have deduced arithmetic from logic. For my argument has been simply that it is part of this claim that the propositions we deduce are indeed the propositions we are already familiar with in arithmetic; the logicist must ensure that he has established this point, and what we are now doing is setting down as axioms the premisses that he needs to establish it.

It should be observed that the axioms by which our important definitions are replaced are not themselves claims to sameness of meaning; the axioms are simply the claims to equivalence that they appear to be. For the procedure I am now advocating consists in always understanding a new formula to have been created with the meaning we desire it to have as a representation of the already familiar concept, and then setting down an axiom for the new formula which asserts that if this is its meaning then it is equivalent to the formula previously given as its definiens. The original arguments given for the 'adequacy' of our definitions are now seen much more appropriately as arguments offered to establish the *truth* of the new axioms, for the truth of these axioms is something that we do need to establish. What arithmetic is deduced *from* can then be seen as the original 'purely logical' basis together with the additional axioms corresponding to our old definitions.

But if such additional axioms are needed for the success of the deduction, what has become of the thesis of logicism? All that has happened, I think, is that we have destroyed that version of logicism which claims that arithmetic simply *is* logic, for to establish this claim it would be necessary to show that the additional axioms are still 'purely logical' axioms, and there seems to be no sense to such a contention. But of course Frege himself did not put his claim in this way at all: he maintained rather that the truths of arithmetic were *analytic* truths, and this claim it seems to me that we can very properly endorse. For on any usual understanding of the concept of analyticity a proposition is shown to be an analytical truth if it is proved, by purely logical reasoning, from premisses that are themselves analytical truths; and it seems to me that the premisses required for our deductions do merit this title, for in each case their justification is that they can be seen to follow—some more and some less immediately—from an *analysis* of the relevant concept. But whether or not one takes this as a complete vindication of the philosophical thesis of logicism, at least it does do something to clarify the status of the truths of arithmetic, and so to provide a solution to one ancient problem.

But another still remains, namely the *meaning* of these truths.

3. The Analysis of Arithmetical Propositions

The argument of my last section was designed to show that since it is a part of the logicist's thesis that the propositions he has deduced are the familiar propositions of arithmetic, he has to establish that the formulae he is employing do mean what the familiar formulae of arithmetic mean. I suggested, however, that he would do best not to argue for this point by persisting in the view that his formulae have exactly the meaning given to them by his definitions, and then going on to claim that this just is the meaning of our familiar expressions, for at least in the case of the ancestral this claim appears to be quite unreasonable. What he can do, however, is to stipulate from the start that his new formulae *are* to be understood as having the familiar meaning, so that his definitions are now transformed into axioms claiming that an equivalence holds, and the status of the truths of arithmetic—i.e. whether they count as 'logical' truths or 'analytic' truths or whatever—will then be the same as the status of these axioms. But perhaps, once the detailed justifications of the axioms have actually been given, it is no longer very important to decide what status-label to attach to them.

Now it is chiefly the case of the ancestral that forces us to adopt this procedure, because there does appear to be a very notable difference in the meaning of definiens and definiendum in this case. With regard to the other definitions just discussed, namely those of '$(\exists n\alpha)(\Phi\alpha)$' and '$(\exists n_{\mathfrak{R}}\alpha)(\Phi\alpha)$', although I did point out some advantages to be gained by treating them explicitly as axioms, still I would also think it quite reasonable to claim a sameness of meaning here, at least to the extent that this is properly required for an analysis.[13] One might perhaps hold, then, that of the premisses required to establish the logic of numerical quantifiers, and to show that it *is* the logic of numerical quantifiers, it is only that concerning the ancestral that occasions any real difficulty. But so far I have not offered any discussion of the move from formulae involving numerical quantifiers to the genuinely arithmetical formulae employing the relations '$\ldots =_N \ldots$' and '$\ldots \mathbf{S}_N \ldots$', and this gap should now be filled.

It is a very simple matter to show that the definitions I offered here are 'adequate' in the same sense as is the definition of the ancestral. For if we now deliberately understand the formula '$n =_N m$' to mean what is meant by 'n is the same number as m', then whatever exactly this does mean it is at least obvious that if n and m are the same number the associated numerical quantifiers must be equivalent, and so we can set down as an obvious truth

$$\vdash n =_N m \supset (\forall \mathfrak{R})(\exists n_{\mathfrak{R}} \equiv \exists m_{\mathfrak{R}}).$$

But it is equally obvious that if there are *exactly* n so-and-so's and at the same time *exactly* m so-and-so's then the two numbers n and m must be equal, i.e. they must be same number; rendering the strong numerical quantifiers just employed in terms of the more familiar weak quantifiers, we may therefore also set it down as obvious that

$$\vdash (\exists \mathfrak{R})(\exists \phi)((\exists n_{\mathfrak{R}}\alpha)(\phi\alpha) \;\&\; \sim(\exists n'_{\mathfrak{R}}\alpha)(\phi\alpha) \;\&\; (\exists m_{\mathfrak{R}}\alpha)(\phi\alpha)$$
$$\&\; \sim(\exists m'_{\mathfrak{R}}\alpha)(\phi\alpha)) \supset n =_N m$$

and now it will be a simple matter to deduce, from our two 'obvious truths', that

$$\vdash n =_N m \equiv (\forall \mathfrak{R})(\exists n_{\mathfrak{R}} \equiv \exists m_{\mathfrak{R}}).$$

The *truth* of this equivalence is all that we need to be granted if we are concerned to construct the logic of arithmetical proposi-

[13] But with '$(\exists n_{\mathfrak{R}}\alpha)(\Phi\alpha)$' we should limit this claim to cases where the condition 'Count ($\mathfrak{R}$)' is satisfied, for the definition is admittedly arbitrary in other cases.

tions from the logic of propositions containing numerical quantifiers, and its truth does follow from the two very obvious truths just given. But I think one will hardly be satisfied to leave the matter at that; for whereas in the previous cases the equivalences we required were at least justified on the basis of an analysis of the concepts involved, in this case we cannot really be said to have produced any analysis at all.

The awkwardness of the situation will perhaps become clearer if we note that the equivalence just argued for is by no means the only equivalence that will satisfy our requirements. For instance, I have already mentioned that it would be equally correct to assert

$$\vdash n =_N m \equiv (\exists n_{\neq} \equiv \exists m_{\neq})$$

and indeed

$$\vdash n =_N m \equiv (\exists n_{\not\equiv} \equiv \exists m_{\not\equiv})$$

and the same holds for many other counting relations. Again it is easily shown from the deduction of Peano's postulates that

$$\vdash n =_N m \equiv (\exists n''_{\neq} \equiv \exists m''_{\neq})$$

while it is simple enough to produce a proof that

$$\vdash n =_N m \equiv (\forall q)(\exists n_{\neq} * \mathbf{S}q \equiv \exists m_{\neq} * \mathbf{S}q)$$

and further examples could be produced without trouble. But it would not seem that all these various equivalents can easily be viewed as having the same meaning as one another, and therefore we must apparently ask *which* of them (if any) has the same meaning as the numerical identity we are bent on defining. And how is this question to be answered?

With the difficulty developed only thus far, I think one could quite plausibly reply that the equivalence we first selected has the best claim of those so far mentioned. For, as a rough attempt, one might first suggest that to say that n is the same number as m is to say that the one may always be substituted for the other *salva veritate*, and this generality is not well expressed if we concentrate attention upon one particular counting relation—as in '$(\exists n_{\neq} \equiv \exists m_{\neq})$' or '$(\exists n_{\not\equiv} \equiv \exists m_{\not\equiv})$'—rather than speaking generally of them all. On the other hand it is not in fact quite true that if n and m are the same number then the one may be substituted for the other in *all* contexts, but only in such contexts as are extensional, and this is very much what is suggested by the equivalence I have singled out. For if we suppose for a moment that (except perhaps for their occurrence in arithmetical propositions) numbers occur only as elements in our numerical quantifiers, then we may say that this equivalence expresses the fact that any quantifiers associated with n and m, whatever their counting relations, will be intersubstitutible in any proposition in which they have *primary* occurrence—i.e. in any propositions which contain those quantifiers *initially*. But it is a perfectly general feature of the logic of quantifiers that quantifiers which are intersubstitutible in initial position are also intersubstitutible in any other extensional position, though the converse is not generally true. For instance the example '$(\exists n''_{\neq} \equiv \exists m''_{\neq})$' concerns the intersubstitutibility of the quantifiers '$\exists n_{\neq}$' and '$\exists m_{\neq}$' in one particular non-initial position, while the example '$(\forall q)(\exists n_{\neq} {*}Sq \equiv \exists m_{\neq} {*}Sq)$' concerns propositions which contain those same quantifiers in another non-initial position,[14] and although it *can* be shown that since the quantifiers are intersubstitutible in these particular non-initial positions they must also be intersubstitutible in initial position too, the argument is not altogether a trivial one. In the other direction, however, the argument is quite immediate, and so it does seem more natural to single out the initial position if what one is really concerned with is all extensional positions. Besides, there is an important and suggestive feature of the definition

$$\text{'}n =_N m\text{'} \quad \text{for} \quad \text{'}(\forall \mathfrak{R})(\exists n_{\mathfrak{R}} \equiv \exists m_{\mathfrak{R}})\text{'}$$

which it would be a pity to lose: this definition very easily lends itself to the suggestion that the concept of a number is something abstracted from the concept of a numerical quantifier simply by discounting the counting relation; and there is surely *some* truth in that.

But on closer reflection, it is not the whole truth; for we have not by any means resolved the original difficulty that we may easily construct very many different formulae that are all equivalent to '$n =_N m$'. Perhaps we might allow that this suggested definition does adequately capture the intersubstitutibility in all extensional contexts of the weak numerical quantifiers '$\exists n_{\mathfrak{R}}$' and '$\exists m_{\mathfrak{R}}$'. But now, why have we confined ourselves to these *weak* quantifiers? For clearly, numbers also

[14] Non-initial, because of the definitions of '$\exists n'$' and '$*S$'.

occur as elements in *strong* numerical quantifiers as well as weak ones, and we appear to be losing sight of that fact.

The series of strong numerical quantifiers I initially indicated by the string of definitions

$$\text{‘}(0x)(Fx)\text{’ for ‘}{\sim}(\exists x)(Fx)\text{’}$$
$$\text{‘}(1x)(Fx)\text{’ for ‘}(\exists x)(Fx \,\&\, (0y)(Fy \,\&\, y \neq x))\text{’}$$
$$\text{‘}(2x)(Fx)\text{’ for ‘}(\exists x)(Fx \,\&\, (1y)(Fy \,\&\, y \neq x))\text{’}$$

etc.

By considerations exactly parallel to those advanced in the last chapter we could evidently ‘define’ the general form ‘$(n\alpha)(\Phi\alpha)$’ by putting

$$\text{‘}\vdash (\ldots(n\alpha)(-\alpha-)\ldots)\text{’ for ‘}\vdash (Q{*}S0 \supset (\ldots(Q\alpha)(-\alpha-)\ldots))\text{’}$$

(and similarly for numerical variables), and the logic of the strong numerical quantifiers based on non-identity would then fall out as a consequence. As I initially remarked, the deductions prove to be more complex in this case, but since they can perfectly well be given there is no very compelling reason for preferring the one set of quantifiers to the other.[15] Indeed either set of quantifiers may be defined in terms of the other in view of the equivalences

$$\vdash (n\alpha)(\Phi\alpha) \equiv ((\exists n\alpha)(\Phi\alpha) \,\&\, {\sim}(\exists n'\alpha)(\Phi\alpha))$$
$$\vdash (\exists n\alpha)(\Phi\alpha) \equiv (\forall \mathfrak{m})(n{*}S\mathfrak{m} \supset {\sim}(\mathfrak{m}\alpha)(\Phi\alpha))$$

which we could prove (except of course for $n =_N 0$) if the two sets of quantifiers were introduced independently of one another.

[15] A deduction based on the strong quantifiers is given by Borkowski who, however, employs, a different definition of the successor relation. If we retain my definition of this relation then in order to obtain analogues for my theorems 1 and 2 we need to alter the definition of ‘$\ldots \geqslant_N \ldots$’ by putting

$$\text{‘}1\text{–}1\mathfrak{R}\text{’ for ‘}(\forall\alpha\beta\gamma)((\alpha\mathfrak{R}\beta \,\&\, \alpha\mathfrak{R}\gamma \supset \beta=\gamma) \,\&\, (\alpha\mathfrak{R}\beta \,\&\, \gamma\mathfrak{R}\beta \supset \alpha=\gamma))\text{’}$$
$$\text{‘}\Phi \text{ sm } \Psi\text{’ for ‘}(\exists\mathfrak{R})(1\text{–}1\mathfrak{R} \,\&\, (\forall\alpha)(\Phi\alpha \supset (\exists\beta)(\Psi\beta \,\&\, \alpha\mathfrak{R}\beta))$$
$$\&\, (\forall\beta)(\Psi\beta \supset (\exists\alpha)(\Phi\alpha \,\&\, \alpha\mathfrak{R}\beta)))\text{’}$$
$$\text{‘}n \geqslant_N m\text{’ for ‘}(\forall\phi\psi)((n\alpha)(\phi\alpha) \,\&\, (m\alpha)(\psi\alpha) \supset (\exists\chi)((\forall\alpha)(\chi\alpha \supset \phi\alpha) \,\&\, \chi \text{ sm } \psi))\text{’}.$$

With this new definition theorems 1 and 2 may indeed be re-established (with theorem 1 now in the form ‘$\vdash n \geqslant_N 0$’), but evidently only after some labour. Theorem 6 then follows as before, but the chief use of theorem 6 depended on the fact that

$$\vdash (n \geqslant_N m \,\&\, m \geqslant_N n) \supset n =_N m$$

and this, which used to be immediate, will now require us to prove the Schröder–Bernstein theorem.

Consequently there would not appear to be any very satisfactory reason for regarding the weak quantifiers as in some way more fundamental than the strong ones, and so it would appear that the definition[16]

$$\text{'}n =_N m\text{'} \quad \text{for} \quad \text{'}(\forall \mathfrak{R})(n_{\mathfrak{R}} \equiv m_{\mathfrak{R}})\text{'}$$

has just as good a claim to yield the meaning of a numerical identity as does our first suggestion

$$\text{'}n =_N m\text{'} \quad \text{for} \quad \text{'}(\forall \mathfrak{R})(\exists n_{\mathfrak{R}} \equiv \exists m_{\mathfrak{R}})\text{'}$$

—yet again the meaning of the definiens would not appear to be the same in both cases. (In fact the definiens with weak quantifiers immediately implies the definiens with strong quantifiers, in view of the above definition of strong quantifiers in terms of weak ones; but the converse argument is altogether more complicated.) So, to revert to the earlier way of putting things, we are apparently faced with the question: is the concept of a number abstracted from the concept of a *weak* numerical quantifier or from the concept of a *strong* numerical quantifier? For it is not altogether easy to see how we may reply: both.

Yet the problem does not stop here, as we may see by trying out an answer to what has been said so far. For suppose we try saying that there is indeed a number 3 which is abstracted from the weak quantifiers '$\exists 3_{\mathfrak{R}}$', and there is also a number 3 which is abstracted from the strong quantifiers '$3_{\mathfrak{R}}$' but it is *not the same number 3*. Or, more strictly, suppose we say that the two formulae '$(\forall \mathfrak{R})(\exists n_{\mathfrak{R}} \equiv \exists m_{\mathfrak{R}})$' and '$(\forall \mathfrak{R})(n_{\mathfrak{R}} \equiv m_{\mathfrak{R}})$' do each give the sense of a claim that n is the same number as m, only it is not the same claim in the two cases, and the ordinary arithmetical formula '$n = m$' is ambiguous between them. The idea would then be that '$n = m$' in its ordinary arithmetical usage would be ambiguous between these two claims in just the way that it is also ambiguous between either of these and what is more strictly expressed as '$+n = +m$', or again what is more strictly expressed as '$n/1 = m/1$'. For, just as it is common logicist practice to distinguish the rational number 3/1 from the signed

[16] I take it that the strong quantifiers are so defined that (except where $n =_N 0$)

$$\vdash (n_{\mathfrak{R}}\alpha)(\Phi\alpha) \equiv ((\exists n_{\mathfrak{R}}\alpha)(\Phi\alpha) \;\&\; {\sim}(\exists n'_{\mathfrak{R}}\alpha)(\Phi\alpha)).$$

But this definition will be discussed more fully in the opening chapter of Part II of this book.

integer $+3$, and these again from the natural number 3, so now we should be invoking a further distinction *within* what are usually called the natural numbers, which we might mark by retaining '3' as the official symbol for what we will call a *cardinal* number associated with the strong quantifiers '$3_{\mathfrak{R}}$', but using '$\exists 3$' as the official symbol for what we will call a *counting* number associated with the weak quantifiers '$\exists 3_{\mathfrak{R}}$'. The differences between the natural numbers on the one hand and the full systems of signed integers or rational numbers on the other, are certainly very much greater than the differences between the natural cardinals and the counting numbers. Most notably, whereas both the natural cardinals and the counting numbers constitute *progressions*, with a first number and a next and a next and so on in order, the signed integers differ in having no *first* number while the rationals differ in having no *next* numbers. By contrast with this the corresponding distinction between the natural cardinals and the counting numbers is altogether trivial, consisting simply in the fact that the natural cardinals begin with 0 while the counting numbers begin with 1. But this is still a sufficient distinction for us to say that the arithmetic of natural cardinals contains the arithmetic of counting numbers as a proper fragment, just as the arithmetic of signed integers or of rational numbers contains the arithmetic of natural cardinals as a proper fragment. Furthermore, there is a genuine difference in the function of the two varieties of natural numbers, as I have indicated in my nomenclature; for the procedure of counting starts, of course, with 1 and not with 0, and it is entirely plausible to view counting as depending upon the weak numerical quantifiers throughout, in so far as one counts 'at least 1, at least 2, at least 3,... at least n; and no more'. The result reached by this procedure will then be summed up with the help of the strong quantifiers, and indeed most answers to the question 'how many?' will be given with these quantifiers, but one possible answer to the question is that there are none to be counted anyway, and this answer employs a natural cardinal though it is clearly not reached by a counting procedure.

So, at first sight it would appear that we could answer our problem by distinguishing two different varieties of natural numbers, and indeed I would maintain that it is just *as* reasonable to recognize a distinction between the natural cardinal 3 and the counting number 3 as it is to recognize a distinction between each of these and, say, the positive integer 3. But really it is not at all clear that the distinction is a reasonable one in *any* of these cases. The idea behind such distinctions is that we should be able to take an ordinary arithmetical formula, such as '$3 > 2$', as ambiguous between the distinct propositions which we can express more precisely in the distinct formulae

$$\exists 3 > \exists 2$$

$$3 > 2$$

$$+3 > +2$$

$$3/1 > 2/1.$$

But a natural English rendering for these formulae would be something like[17]

At least 3 is more than at least 2

Exactly 3 is more than exactly 2

3 more is more than 2 more

3 times as much is more than 2 times as much.

And the difference between these four propositions seems to be that they each employ *the same* numbers 3 and 2, but each in a different *special context*. Certainly one is very likely to suppose that the original formula '$3 > 2$' with its simple English rendering '3 is more than 2', is not so much ambiguous between these four propositions (and others like them) but rather a more fundamental truth that underlies them all. Or again, to revert to the problem of numerical identity that we began with, it seems much more reasonable to hold that when we say that n is the same number as m we are in this one proposition implying that either may be substituted for the other, not only in the contexts 'there are at least n' and 'there are exactly n', but also in the further contexts 'n more' and 'n times as much' which we have not yet even discussed. For it simply is not true to say that (except perhaps for their occurrence in arithmetical propositions) numbers occur only as elements of (strong or weak) numerical

[17] The rendering of the fourth formula may seem a little unexpected, but I am drawing here on the second part of this book.

quantifiers. But this must have the consequence that an attempt to define numerical formulae in terms of formulae involving numerical quantifiers is bound to appear somewhat arbitrary and unconvincing. To achieve a proper understanding of the sense of arithmetical formulae we should at least survey more widely the various different contexts in which numbers appear, and then we may reconsider the question whether the formulae of arithmetic can indeed be understood as mere generalizations over all these various different contexts.

This brings me to the subject of the second part of this book, in which I shall consider the natural numbers more widely, and investigate their relations to the so-called rational and irrational numbers. Meanwhile, I bring the present part to an end with a summary of my discussion of logicism in this chapter.

My general argument has been that although the logicist programme can indeed be carried through, nevertheless the significance of this fact for our understanding of the notion of number is not as great as might have appeared. In order to carry through the logicist programme one must start with a logical basis rather different from that pioneered by Frege, and in order to forestall criticism stemming from the incompleteability of formal arithmetic one must not tie one's conception of logic too closely to the ideal of a fully formalized deductive system, but neither of these points seems to me at all a serious setback. By making very full use of contextual definition it is then possible to prove in one's logic formulae that may be abbreviated, by the definitions given, to other formulae that correspond exactly to the usual theorems of arithmetic; but since the logicist claim is not just that these formulae *correspond to* the truths of arithmetic but that they do actually express those truths, there is still much more to be said. In particular it has to be claimed that the definitions in question either are or (better) are justified by correct analyses of the relevant concepts, and *this* claim is not one that can be settled by 'purely logical' argument. In fact the claim is not easy to assess, and we found some difficulties with it even before we came to the analysis of arithmetical propositions and were still concerned with the introduction of numerical quantifiers. But with the analysis of arithmetical propositions the difficulty seems particularly acute.

Since the properties of the numerical quantifiers do correspond exactly to the properties commonly said to be the properties of natural numbers, it would perhaps have been possible to maintain that the propositions of arithmetic commonly said to be about natural numbers simply *are* the corresponding generalizations over numerical quantifiers, but only if we were also prepared to acquiesce in the traditional view that the natural numbers should be distinguished from the positive integers which bear the same name, and similarly from the integral rationals, and so on. (This would make it appropriate to distinguish within the natural numbers a series of numbers beginning with 0 and another series of different numbers beginning with 1, but such a distinction seems at least as reasonable as the usual ones.) However, I think that we should *not* rest content with these traditional distinctions, and therefore it must now appear arbitrary to limit our attention to the numerical quantifiers to the exclusion of other types of numerical expression—for instance the ordinal expressions 'is the n'th' or the comparative expressions 'is n times as . . . as' which will be discussed in the next part of the book. Looking at the matter from a Platonist viewpoint, one might perhaps say that it is a mistake to think of cardinal numbers, ordinal numbers, rational numbers, and so on, as different kinds of number; rather, there are cardinal *uses* of numbers, and ordinal uses, and uses in numerical comparisons, and still other uses; but the numbers themselves should be distinguished from these several uses of them. But whether we too shall in the end have to adopt this Platonist viewpoint must await further investigation.

List of Works Cited

BENACERRAF, PAUL. 'What Numbers Could Not Be', *Philosophical Review*, lxxiv (1965), 47–73.

BORKOWSKI, LUDWIK. 'Reduction of Arithmetic to Logic based on the Theory of Types without the Axiom of Infinity and the Typical Ambiguity of Arithmetical Constants', *Studia Logica*, viii (1958), 283–95.

CHURCH, ALONZO. *Introduction to Mathematical Logic*, i, Princeton University Press, Princeton, 1956.

—— 'A Formulation of the Simple Theory of Types', *Journal of Symbolic Logic*, v (1940), 56–68.

FREGE, GOTTLOB. *Grundlagen der Arithmetik*, Breslau, 1884. Reprinted with an English translation by J. L. Austin as *The Foundations of Arithmetic*, Blackwell, Oxford, 1950.

—— *Grundgesetze der Arithmetik*, Jena, 1893 and 1903. Partially translated by M. Furth as *The Basic Laws of Arithmetic*, University of California Press, Berkeley and Los Angeles, 1964.

FURTH, MONTGOMERY. See above, under Frege.

GEACH, PETER T. 'Class and Concept', *Philosophical Review*, lxiv (1955), 561–70.

—— *Reference and Generality*. Cornell University Press, New York, 1962.

GODDARD, L. 'Sense and Nonsense', *Mind*, lxiii (1964), 309–31.

—— and ROUTLEY, R. 'Use, Mention, and Quotation', *Australasian Journal of Philosophy*, 44 (1966), 1–49.

GOODMAN, NELSON, and QUINE, W. V. O. 'Elimination of Extra-Logical Postulates', *Journal of Symbolic Logic*, v (1940), 104–9.

KNEALE, W. C., and KNEALE, M. *The Development of Logic*, Clarendon Press, Oxford, 1962.

LEMMON, E. J. *Beginning Logic*, Nelson, London and Edinburgh, 1965.

PAP, ARTHUR. 'Types and Meaninglessness', *Mind*, lxix (1960), 41–54.

PARSONS, C. D. 'Frege's Theory of Number', in *Philosophy in America*, ed. M. Black, George Allen and Unwin, London, 1964, pp. 180–203.

PUTNAM, HILARY. 'The Thesis that Mathematics is Logic', in *Bertrand Russell, Philosopher of the Century*, ed. R. Schoenman, George Allen and Unwin, London, 1967, pp. 273–303.

QUINE, W. V. O. *Mathematical Logic*, New York, 1940. Revised edn., Harvard University Press, Cambridge, Mass., 1951.

—— 'Notes on Existence and Necessity', *Journal of Philosophy*, 40 (1943), 113–27.

QUINE, W. V. O. *From a Logical Point of View.* Harvard University Press, Cambridge, Mass., 1953.
—— *Methods of Logic*, New York, 1950. Revised edn., Routledge and Kegan Paul, London, 1962.
—— *Word and Object*, M.I.T. Press, Cambridge, Mass., 1960.
—— and GOODMAN, NELSON. See above, under Goodman.
RAMSEY, F. P. *The Foundations of Mathematics and other Logical Essays*, ed. R. B. Braithwaite, Routledge and Kegan Paul, London, 1931.
ROUTLEY, R., GODDARD, L. See above, under Goddard.
RUSSELL, BERTRAND. *The Principles of Mathematics*, Cambridge University Press, Cambridge, 1903.
—— *Introduction to Mathematical Philosophy*, George Allen and Unwin, London, 1919.
—— *Logic and Knowledge, Essays 1901–1950*, ed. R. C. Marsh, George Allen and Unwin, London, 1956.
—— and WHITEHEAD, A. N. *Principia Mathematica* i, Cambridge University Press, Cambridge, 1910.
WHITEHEAD, A. N., and RUSSELL, BERTRAND. See above, under Russell.

Index

LIBRARY
OF
MOUNT ST. MARY'S
COLLEGE
EMMITSBURG, MARYLAND